ELECTRIC WAVES

ELECTRIC WAVES

BEING AN ADAMS PRIZE ESSAY IN THE UNIVERSITY OF CAMBRIDGE

BY

H. M. MACDONALD, M.A., F.R.S.

FELLOW OF CLARE COLLEGE, CAMBRIDGE.

CAMBRIDGE:

AT THE UNIVERSITY PRESS.

1902

CAMBRIDGE
UNIVERSITY PRESS

University Printing House, Cambridge CB2 8BS, United Kingdom

Published in the United States of America by Cambridge University Press, New York

Cambridge University Press is part of the University of Cambridge.

It furthers the University's mission by disseminating knowledge in the pursuit of education, learning and research at the highest international levels of excellence.

www.cambridge.org
Information on this title: www.cambridge.org/9781107695191

© Cambridge University Press 1902

First published 1902
First paperback edition 2013

A catalogue record for this publication is available from the British Library

ISBN 978-1-107-69519-1 Paperback

PREFACE.

THE following essay was undertaken with the object of discussing the possibility of obtaining directly from Faraday's laws a consistent scheme for the representation of electrical phenomena, and of applying the results to obtain the quantitative relations which exist in certain cases of the propagation of electrical effects.

Maxwell's memoir on "A dynamical theory of the Electromagnetic Field," communicated to the Royal Society in October, 1864, marks a new departure in electrical theory. In it the analytical representation of Faraday's laws is systematically developed and applied, as also the analytical formulation of the electromagnetic theory of light which had already been proposed by Faraday in 1846; but these contributions to electrical theory, though of great importance, are subservient to the main object of the paper, which is to shew that the laws of electrical phenomena obey the same general principle as the laws of mechanics. It has not been sufficiently noticed that Maxwell presented his theory under two forms, in one of which the electrokinetic energy is expressed in terms of currents, and in the other in terms of magnetic force and induction. The first form is the one used throughout the paper, the second form being given without application; and in his Treatise the first form is used in the discussion of the general theory, the second form being given later and only applied to discuss the pressure of radiation and magneto-optic relations. Subsequent writers, FitzGerald, Heaviside, Hertz and others, have taken the second form of the energy function as the starting-point of their investigations.

The fact, that in certain cases the direct application of Faraday's laws gives without ambiguity results different from those which appear to follow from the latter form of Maxwell's theory, led the writer to suspect that there must be some flaw in the process of reasoning by which this form of Maxwell's theory is deduced from Faraday's laws. This suggested the procedure adopted in this essay, to begin by applying Faraday's laws, without the intervention of any dynamical theory, to the different cases which can arise, and to examine whether the results obtained are consistent with observation. A satisfactory scheme having been developed in this way, a short account of Maxwell's procedure is given with the view of discovering the source of the discrepancy, and the result of examination is to shew that the first form in which Maxwell presented his theory is a logical consequence of Faraday's laws while the second form is not. That this has not been noticed earlier is to be explained by the fact that Maxwell did not use the second form of his theory to obtain results whereas subsequent writers have used it in preference to the first form, and that, so far as the applications made by Maxwell are concerned, the same results follow from either form of the theory. That the second form of the theory is easier to work with is obvious, as in this form the Lagrangian function is of the same type as the Lagrangian function of a material medium, thus allowing the argument from analogy to be used, and this explains the preference shewn for this form of the theory.

Having found that the second form of the theory is not logically involved in the first, it becomes necessary to ascertain

what assumptions its use has led to, and it appears that many
of the received ideas as to the nature of the aether, as for
example the doctrine of a fixed aether, have arisen in this way.
These assumptions not being directly concerned in the first
form of Maxwell's theory, and this form being the one which
logically follows from Faraday's laws, the ideas that have arisen
from the assumptions cannot be regarded as being required by
the facts.

The next step is to examine whether the form of Maxwell's
theory thus adopted is consistent with the laws of dynamics,
and the logical development of this form of the theory is then
resumed at the point where it was left by Maxwell. This
latter is perhaps unnecessary in view of the fact that earlier
in the essay it is shewn that Faraday's laws are sufficient in
themselves for the development of a scheme of representation ;
but it was thought desirable to add it for the sake of com-
pleteness.

The object of the second part of the essay is the application
of the general theory to some of the problems that present
themselves in connection with the propagation of electrical
effects. It appears that there is an essential difference between
a simply-connected and a multiply-connected space in respect of
the propagation of electrical effects, there being no permanent
free oscillations such as would not be dissipated by radiation belonging to an indefinitely extended simply-connected space, while
there are such permanent free oscillations associated with each
of the conducting circuits that render an indefinitely extended
space a multiply-connected region. The explicit recognition of
this fact makes it possible to simplify the mathematical theory
of electric waves. The complete determination of the circumstances of propagation of waves of the latter class can be reduced
to the solution of linear differential equations involving one
independent variable : the dependent variable belonging to any

conducting path returns to the same value on going once round
the path when closed, and the waves circulate without decay.
In an open path the dependent variable vanishes at both ends
and the energy is dissipated in radiation. The principal application of the theory is to the effects investigated by Hertz in his
experiments on electric waves. The development of the analysis
gives results which are in close agreement with the observations
of Sarasin and de la Rive, who repeated Hertz's experiments
under more favourable conditions; it also gives a satisfactory
explanation of the discrepancies in some of Hertz's own observations. The investigation thus given for the free periods
of resonators is only approximate; the accurate equation for
the determination of the periods of a resonator can be easily
deduced from the general theory of these waves, but this
equation is such that the calculation of the periods from it
would involve great labour, and as the error involved in using
the approximate theory, in which the distance between successive nodes is approximately half a wave-length, must be
extremely small, the gain in theoretical accuracy did not appear
to be sufficient compensation for the additional labour.

The essay in its present form was completed at the end of
1900, with the exception of Chapter VIII, the paragraph
relating to it in Chapter I, and the Appendices. The para-
graphs have also been renumbered throughout and the cross
references altered accordingly. Chapter VIII has been added
with the view of developing the manner in which the energy of
permanent vibrations associated with closed circuits is distributed, and the distinction between them and the waves due to
open oscillators. To effect this it was found desirable for the
sake of uniform treatment to give a short account of radiation,
starting from the first form of Maxwell's theory. Appendices
A and B are intended to elucidate the point of view adopted in
the first part of the essay. Appendix C supplies the analysis

belonging to Chapter V, which was omitted on the ground
that it might unduly interfere with the course of the argument.
Appendix D contains an application of a method developed in
the essay to a dynamical problem in diffraction which includes
all the cases hitherto solved; this appendix also contains the
analytical investigation of a result stated without proof in
Chapter VI.

In the theoretical discussion references are given to other
writings only when these bear directly on the argument, this
being thought sufficient in an essay. On the other hand, where
special problems are dealt with, references have, as far as is
known, been given to all the previous investigations.

To Mr J. Larmor and to Prof. A. E. H. Love, whose assist-
ance in revising the sheets for the press he was fortunate enough
to obtain, the writer is greatly indebted. The many valuable
criticisms and suggestions received from them have done much
to remove obscurities and improve the book in other ways.

Acknowledgement is due to the officials of the University
Press for the careful and obliging manner in which they have
done their share of the work.

CLARE COLLEGE, CAMBRIDGE.
August 4, 1902.

CONTENTS.

CHAPTER I.

INTRODUCTION.

1. THE object of the following essay is the discussion of
the circumstances of propagation of electrical effects. The first
part consisting of the second, third, fourth and fifth chapters
is devoted to the discussion of the general theory. The
equations of electrodynamics for a single medium in which
there are conductors at rest relatively to each other are deduced
from Faraday's laws, the current at a point being, in accordance
with Maxwell's views, supposed to be made up of two parts,
the aethereal displacement current and the convection current
at that point. The integrals of these equations, when waves
of definite period are being propagated in the medium, are
obtained in terms of the distribution of the convection currents
throughout the space; the electric and magnetic forces at each
point of the medium, which belong to waves of the period
chosen, are then expressed in terms of the distribution of
convection currents having the same period. The expressions
in the general case, where there are waves of different periods,
are sums of these. These integrals are then modified to suit
the case of perfect conductors, and it appears that, in this case,
the expressions for the electric and magnetic forces at each
point of the medium take the form of integrals over the
surface of the conductors, the unknown quantity being the
magnetic force at the surface of a conductor. It is a consequence of these results that, when the components of the
magnetic and electric forces tangential to a surface at each
point of it are known, this surface being closed and enclosing

all the sources of the waves, the electric and magnetic forces at any point can be immediately deduced. The propagation of any disturbance through the medium is shortly discussed, and it follows from these investigations that the way, in which Maxwell's aethereal displacement is introduced, involves the assumption that it, as well as the electric force conceived of as belonging to it, are independent of the motion of the aether.

2. In the third chapter convection currents in motion and material media are considered, the difference between material media and the aether being taken to be the presence of convection currents. Faraday's laws are applied on this assumption to obtain the equations of propagation of electric waves in transparent media, and the results so arrived at are in agreement with Fresnel's formula for the effect of moving material media. The application of these laws to any material medium leads to equations determining the circumstances and to expressions for the mechanical forces acting on material media which are the same as those used by Larmor, *Phil. Trans.* A. 1897.

3. Maxwell's dynamical theory of electricity is discussed in the fourth chapter, where the manner in which the various functions arise is considered and in particular it is shewn that on his theory aethereal displacement is a vector which defines the electrical degrees of freedom of the aether at each point of space, this vector being conceived of as associated with the points of space not with definite elements of the aether. It is then observed that the Lagrangian function of the motions of the aether which is obtained on this theory is a modified Lagrangian function, the coordinates specifying degrees of freedom of the aether other than electrical having been eliminated by the process by which the function has been built up; so that although this function, when the convection currents are completely known, is sufficient to determine the electrical relations of the aether to material media, it does not supply sufficient knowledge from which to develop the dynamics of the aether. Maxwell's second expression for the electrokinetic energy, which expresses it in terms of the magnetic force and the magnetic induction, is shewn to have been obtained from the previous one by a transformation which is in general invalid. It then appears that it is the use of this second expression which has led to the assumption that Maxwell's electromagnetic theory is the same as MacCullagh's optical theory and thence to the identification of the magnetic induction as the velocity of the aether. This again has given rise to difficulties connected with the effect of a magnetic field on the propagation of light and to difficulties connected with permanent magnets which have compelled the further assumption that the density of the aether is indefinitely great. This second expression for the electrokinetic energy being illegitimate, it follows that Maxwell's theory is essentially different from MacCullagh's and that the conclusions referred to above, which are based on the identification of the two theories, are not a necessary consequence of Maxwell's theory.

4. The conclusion having been arrived at that Maxwell's theory, though sufficient for the treatment of electrical changes did not furnish a complete representation of the motions of the aether, it became necessary to inquire what form a dynamical theory could take when only part of all the possible motions can be taken into account and whether Maxwell's theory is of a form which could arise in this way. This inquiry forms the subject of the fifth chapter; the fundamental assumption made is that if all the degrees of freedom everywhere existing could be put in evidence, the Lagrangian function of all the corresponding motions would be a homogeneous quadratic function of the velocities belonging to all the degrees of freedom, the coefficients in this expression being functions of the coordinates which specify the degrees of freedom, and, further, the motions are assumed to obey the principle of Least Action. The existence of a class of motions, in which the knowledge of a single function, viz. a modified Lagrangian function of the form $T - V$, is sufficient to determine completely the motions corresponding to the degrees of freedom, the coordinates belonging to which have been retained, is demonstrated; the motions belonging to those which have been eliminated cannot be determined from a

knowledge of this function. The expression V, which occurs in this modified Lagrangian function, is equal to the energy of the motions which correspond to the degrees of freedom whose coordinates have been eliminated, so that the potential energy is the energy of the concealed motions. The forms which can arise under other circumstances are also discussed. The Lagrangian function, which is derived from observation, is, in every case, a modified function in which only the observed degrees of freedom appear explicitly. In attempting to apply a dynamical theory to the aether it has to be remembered that the energy of the motions of a continuous medium, such as it must be postulated to be, cannot be represented by a sum of the form $\iiint \rho \, (u^2 + v^2 + w^2) \, dx dy dz$; to do so would be to endow the aether with atomic structure. It then follows that the application of the previous results to the aether, all its possible degrees of freedom being taken into account, can give rise to a modified Lagrangian function the same as that arrived at on the Faraday-Maxwell theory. The manner, in which the effect of material media can be taken into account on this theory, is then discussed and the theory is compared with the other theories which have been proposed.

5. It having been shewn that the Faraday-Maxwell theory gives a consistent account of electrical phenomena and is itself consistent with dynamical theory, and the effect of the motion of an observer and his apparatus having been ascertained, this effect being, in the case of the propagation of electric waves of the kind which occur in experiments such as those of Hertz, negligible, the propagation of electrical effects of this kind is discussed in the second part on this assumption. The propagation of electrical effects in a simply connected space is dealt with in the sixth chapter, where it is shewn that permanent free electrical oscillations in an indefinitely extended simply connected space are impossible. The case of the space between two concentric spherical surfaces is treated in detail in illustration of the general theory. The oscillations belonging to condensers are discussed and the result of an investigation to determine the effect of an open end is given. The effect of joining the opposite faces of a condenser by a thin wire is investigated and it is shewn that, if originally the faces are not closed surfaces, the periods are practically unaltered and no new period is introduced, but that, if originally the faces are closed surfaces, the already existing periods are approximately unaltered and a new period, which is very long in comparison with the others, is introduced. The effect of removing or setting up constraints is discussed, more particularly in the case where the space throughout which the constraint is removed or set up is not finite. Non-permanent free oscillations are discussed, the case of a change in the dielectric medium being treated in detail. The principal result obtained is that reflexion is the essential condition for the existence of free oscillations in a simply connected space.

6. In the seventh chapter the propagation of electrical effects in multiply connected spaces is considered, and it is shewn that permanent free oscillations can exist in a multiply connected space whether the space is infinitely extended or not, the possible free periods for the infinitely extended doubly connected space in which there is a single circuit being s/nV, where s is the length of the circuit, V the velocity of radiation and n any integer. It is then shewn that the problem, in the case of waves travelling along any number of parallel straight cylindrical conductors, is reducible to one of conformal representation. The case of circuits of any form is discussed and expressions for the components of the electric force at any point due to the waves belonging to a single circuit are obtained. The effect of finite cross-section is then considered and it is proved that, taking the cross-section of the wire to be circular, of radius small compared with the radius of curvature of the wire and with the wave-length of the oscillations, the results obtained for the circuit are still applicable. The electric waves induced in a thin closed wire are discussed; and it follows that for any number of thin closed wires the expressions for the components of the electric force at any point are sums of the expressions found for a single closed wire, whence solvable cases for the effect of conductors can be deduced. The waves

belonging to a circuit are unaltered by any others which exist along with it, the effect of conductors being to super-pose on the waves belonging to a circuit waves belonging to other circuits which form the image in the conductors of the first circuit.

7. The radiation of electric energy is treated of in the eighth chapter. In this connexion it was found necessary to investigate the expression for the rate of transfer of energy across a closed surface using Maxwell's first expression for the electrokinetic energy. The result is an addition to Poynting's expression of a part which, integrated throughout a complete period, vanishes so that results, which take account of average radiation only, are unaffected. The intensity of the radiation from, and the density of the distribution of energy due to, a simple oscillator are investigated, and the results are applied to obtain relations between the temperature and the intensity of radiation. The condition of permanence of a group of ions is obtained and the law of the force between permanent groups is deduced, the force between a pair, neither of which possesses free electricity, being at most of the order of the inverse fourth power of their distance apart. It is then shewn that no energy is radiated away from a circuit, as was to be anticipated from the results of the previous chapter.

8. The existence of permanent free oscillations associated with circuits having been established it follows that, if a small gap be made in a circuit, oscillations of this kind, though they will not then be permanent, will be set up in it; the treatment of these open circuits forms the subject of the ninth chapter. It is first proved that, if the case of a straight wire with a free end can be solved, the case of any form of thin wire whose curvature is continuous can be solved, as the waves can only get into the circuit at the open end. The waves set up by the simplest kind of source, which can produce waves of the kind sought in a straight wire with a free end, are then found. The effect of a small sphere placed at the free end is discussed and shewn to be negligible if the radius of the sphere is small compared with the length of the waves. In the tenth chapter

these results are applied to the case of resonators and it appears that the fundamental wave-length of a resonator depends on its length only, the result for a resonator in the form of a circle being that the fundamental wave-length is 7·95 D, where D is the diameter of the resonator, which agrees with the results of the experiments of Sarasin and de la Rive, who obtained the value 8 D for a circular resonator. The rate of decay of the oscillations belonging to a resonator is investigated and shewn to be very small, which agrees with the results of Bjerknes' experiments. The case of a wire with an open end, from which the waves are radiating freely, is then discussed and it is found that, for stationary waves, the distance of the first node from the free end is ·192 λ, where λ is the wave-length of the waves which are being observed, this result again agreeing with the result of the experiments of Sarasin and de la Rive. The forms of the wave-fronts in the neighbourhood of a straight wire, from the end of which the waves are being freely radiated, are then investigated and it appears that, in travelling along the wire from the free end, the wave-fronts change from being para-boloids of revolution with the free end as focus and the wire as axis to being planes perpendicular to the wire; this agrees exactly with the observations of Birkeland and Sarasin. Fur-ther, on the side away from the wire the wave-fronts tend to set themselves at right angles to the radius vector from the free end, which was found to be the case by the same observers. The effect of the upper harmonics of a resonator is considered, and it would seem that they provide an explanation of some results observed by Sarasin and de la Rive and that they are to some extent the cause of the result obtained by Hertz in his interference experiment for comparing the velocity of electric waves in air with their velocity along a wire.

The result of these investigations relative to electric waves in wires and resonators is to shew that there is complete quantitative agreement between experiment and theory as well as qualitative agreement.

CHAPTER II.

THE EQUATIONS OF ELECTRODYNAMICS.

9. THE equations which determine the magnetic and electric forces at each point of space, conceived of as filled by a medium in which there are conductors at rest relatively to each other, can be deduced from Faraday's laws that

(1)   a closed current is equivalent as regards the magnetic field produced by it to a magnetic shell of the same strength bounded by it,

(2)   the electromotive force in a closed circuit is given by

$$E = -\frac{dW}{dt},$$

where W is the number of lines of magnetic induction which pass through it.

Further, on this scheme all currents are regarded as closed.

Denoting the components of magnetic force at any point x, y, z by α, β, γ, the components of magnetic induction by a, b, c, the components of electric force by X, Y, Z and the components of total current strength by u, v, w, the axes of reference being fixed relatively to the conductors, Faraday's laws are expressed by the equations

$$\int \alpha\, dx + \beta\, dy + \gamma\, dz = 4\pi \iint (lu + mv + nw)\, dS \dots (1),$$

$$\int X\, dx + Y\, dy + Z\, dz = -\frac{d}{dt} \iint (la + mb + nc)\, dS \dots (2),$$

where the surface integrals are taken over any surface bounded by an edge, l, m, n are the direction cosines of the normal to the surface at any point and the line integrals are taken round the edge in the positive direction. The relations (1) and (2) are equivalent to the systems of equations

$$
\left.
\begin{aligned}
4\pi u &= \frac{\partial \gamma}{\partial y} - \frac{\partial \beta}{\partial z} \\[2ex]
4\pi v &= \frac{\partial \alpha}{\partial z} - \frac{\partial \gamma}{\partial x} \\[2ex]
4\pi w &= \frac{\partial \beta}{\partial x} - \frac{\partial \alpha}{\partial y}
\end{aligned}
\right\} \dots\dots\dots\dots\dots (1'),
$$

$$
\left.
\begin{aligned}
\frac{da}{dt} &= \frac{\partial Y}{\partial z} - \frac{\partial Z}{\partial y} \\[2ex]
\frac{db}{dt} &= \frac{\partial Z}{\partial x} - \frac{\partial X}{\partial z} \\[2ex]
\frac{dc}{dt} &= \frac{\partial X}{\partial y} - \frac{\partial Y}{\partial x}
\end{aligned}
\right\} \dots\dots\dots\dots\dots (2').
$$

In addition to these equations the relations between magnetic force and magnetic induction and between electric force and current strength must be known in order to effect a solution in any case. For the applications in view the magnetic force may be taken to be everywhere identical with the magnetic induction, so that $a = \alpha$, $b = \beta$, $c = \gamma$. The total current at any point is, according to Maxwell's view, made up of two parts, the displacement current and the convection current. The displacement current has components $\dot{f}$, $\dot{g}$, $\dot{h}$ and is connected with the electric force at the point by the relations

$$
4\pi V^2 f = X, \quad 4\pi V^2 g = Y, \quad 4\pi V^2 h = Z,
$$

where V is the velocity of propagation of electric effects through the medium; the convection current at a point consists of the conduction current at the point, such a current being conceived of as associated with a closed circuit forming a circuital discontinuity in the medium, and an element of current

ρp, ρq, ρr, where p, q, r are the velocities of the free charge of volume density ρ at the point, such a current element being a point discontinuity in the medium and the closed current, of which it must be regarded as forming part, being completed by displacement currents through the medium. The components of total current are then given by

$$\frac{1}{4\pi V^2}\frac{dX}{dt} + u, \quad \frac{1}{4\pi V^2}\frac{dY}{dt} + v, \quad \frac{1}{4\pi V^2}\frac{dZ}{dt} + w,$$

where u, v, w are the components of the convection current at the point x, y, z. In any case the convection current may be regarded as everywhere known or as given by a knowledge of the relation between the electric force and the conduction current, and of the distribution of free charges and their velocities.

10. Before proceeding to discuss the integration of the equations it is convenient to obtain them in a modified form. For this purpose let a vector, whose components are F, G, H and which is connected with the magnetic induction by the relations

$$a = \frac{\partial H}{\partial y} - \frac{\partial G}{\partial z},$$

$$b = \frac{\partial F}{\partial z} - \frac{\partial H}{\partial x},$$

$$c = \frac{\partial G}{\partial x} - \frac{\partial F}{\partial y},$$

be introduced.

Then

$$\frac{da}{dt} = \frac{\partial^2 H}{\partial y\partial t} - \frac{\partial^2 G}{\partial z\partial t},$$

that is

$$\frac{\partial Y}{\partial z} - \frac{\partial Z}{\partial y} = \frac{\partial^2 H}{\partial y\partial t} - \frac{\partial^2 G}{\partial z\partial t},$$

or

$$\frac{\partial}{\partial z}\left(Y + \frac{\partial G}{\partial t}\right) = \frac{\partial}{\partial y}\left(Z + \frac{\partial H}{\partial t}\right),$$

with two similar relations; therefore

$$
\left.
\begin{aligned}
X &= -\frac{\partial F}{\partial t} - \frac{\partial \phi}{\partial x} \\[2mm]
Y &= -\frac{\partial G}{\partial t} - \frac{\partial \phi}{\partial y} \\[2mm]
Z &= -\frac{\partial H}{\partial t} - \frac{\partial \phi}{\partial z}
\end{aligned}
\right\} \quad \dots\dots\dots\dots\dots (3).
$$

Now equations (1') are equivalent to

$$
\frac{1}{V^2}\frac{\partial X}{\partial t} + 4\pi u = -\nabla^2 F + \frac{\partial J}{\partial x},
$$

$$
\frac{1}{V^2}\frac{\partial Y}{\partial t} + 4\pi v = -\nabla^2 G + \frac{\partial J}{\partial y},
$$

$$
\frac{1}{V^2}\frac{\partial Z}{\partial t} + 4\pi w = -\nabla^2 H + \frac{\partial J}{\partial z},
$$

where

$$
J = \frac{\partial F}{\partial x} + \frac{\partial G}{\partial y} + \frac{\partial H}{\partial z},
$$

whence from (3)

$$
\left.
\begin{aligned}
\frac{1}{V^2}\frac{\partial^2 F}{\partial t^2} + \frac{1}{V^2}\frac{\partial^2 \phi}{\partial x \partial t} - 4\pi u &= \nabla^2 F - \frac{\partial J}{\partial x} \\[2mm]
\frac{1}{V^2}\frac{\partial^2 G}{\partial t^2} + \frac{1}{V^2}\frac{\partial^2 \phi}{\partial y \partial t} - 4\pi v &= \nabla^2 G - \frac{\partial J}{\partial y} \\[2mm]
\frac{1}{V^2}\frac{\partial^2 H}{\partial t^2} + \frac{1}{V^2}\frac{\partial^2 \phi}{\partial z \partial t} - 4\pi w &= \nabla^2 H - \frac{\partial J}{\partial z}
\end{aligned}
\right\} \quad \dots\dots(4).
$$

11. Proceeding now to the integration of the equations in the case where an unlimited train of electrical oscillations is being propagated through the medium, those parts only of the electric and magnetic forces which depend on the time need be obtained, and the remaining parts if any will accordingly be omitted. The distribution of the convection currents will be assumed to be known and it will be proved that the electric and magnetic forces at any point can be expressed in terms of these currents. The components F, G, H of the vector introduced above are so far restricted by two independent relations, so that they can be subjected to any other condition not inconsistent with these. Assume as a further condition

$$J = -\frac{1}{V^2}\frac{\partial \phi}{\partial t},$$

where now only those parts of the various quantities involved which depend on the time are considered; then equations (4) become

$$\left.\begin{array}{c} \dfrac{1}{V^2}\dfrac{\partial^2 F}{\partial t^2} - 4\pi u = \nabla^2 F \\[2mm] \dfrac{1}{V^2}\dfrac{\partial^2 G}{\partial t^2} - 4\pi v = \nabla^2 G \\[2mm] \dfrac{1}{V^2}\dfrac{\partial^2 H}{\partial t^2} - 4\pi w = \nabla^2 H \end{array}\right\}\ \ldots\ldots\ldots\ldots(4').$$

Let r denote the distance between the points x, y, z and ξ, η, ζ, then the integrals of equations (4'), which correspond to the propagation of an unlimited train of electrical oscillations of period $2\pi/\kappa V$, are given by

$$F = \iiint \frac{e^{-\iota\kappa(r-Vt)}}{r}\, u_1 d\xi\, d\eta\, d\zeta,$$

$$G = \iiint \frac{e^{-\iota\kappa(r-Vt)}}{r}\, v_1 d\xi\, d\eta\, d\zeta,$$

$$H = \iiint \frac{e^{-\iota\kappa(r-Vt)}}{r}\, w_1 d\xi\, d\eta\, d\zeta,$$

where $u = u_1 e^{\iota\kappa Vt},\quad v = v_1 e^{\iota\kappa Vt},\quad w = w_1 e^{\iota\kappa Vt},$

u, v, w being the components of the convection current at the point ξ, η, ζ and the integrals being taken throughout all space. It follows from these expressions that

$$J = \iiint \left(u_1 \frac{\partial}{\partial x} + v_1 \frac{\partial}{\partial y} + w_1 \frac{\partial}{\partial z}\right)\frac{e^{-\iota\kappa(r-Vt)}}{r}\, d\xi\, d\eta\, d\zeta,$$

whence

$$\frac{\partial \phi}{\partial t} = - V^2 \iiint \left(u_1 \frac{\partial}{\partial x} + v_1 \frac{\partial}{\partial y} + w_1 \frac{\partial}{\partial z} \right) \frac{e^{-\iota\kappa(r-Vt)}}{r} d\xi\, d\eta\, d\zeta,$$

and therefore

$$\phi = \frac{\iota V}{\kappa} \iiint \left(u_1 \frac{\partial}{\partial x} + v_1 \frac{\partial}{\partial y} + w_1 \frac{\partial}{\partial z} \right) \frac{e^{-\iota\kappa(r-Vt)}}{r} d\xi\, d\eta\, d\zeta.$$

Hence by (3)

$$X = - \iota\kappa V \iiint u_1 \frac{e^{-\iota\kappa(r-Vt)}}{r} d\xi\, d\eta\, d\zeta$$

$$- \frac{\iota V}{\kappa} \iiint \left(u_1 \frac{\partial^2}{\partial x^2} + v_1 \frac{\partial^2}{\partial x \partial y} + w_1 \frac{\partial^2}{\partial x \partial z} \right) \frac{e^{-\iota\kappa(r-Vt)}}{r} d\xi\, d\eta\, d\zeta,$$

with two similar relations ; therefore

$$\left. \begin{aligned}
X &= - \frac{\iota V}{\kappa} \iiint \left(u_1 \frac{\partial^2}{\partial x^2} + v_1 \frac{\partial^2}{\partial x \partial y} + w_1 \frac{\partial^2}{\partial x \partial z} + \kappa^2 u_1 \right) \\
&\qquad\qquad \times \frac{e^{-\iota\kappa(r-Vt)}}{r} d\xi\, d\eta\, d\zeta \\
Y &= - \frac{\iota V}{\kappa} \iiint \left(u_1 \frac{\partial^2}{\partial x \partial y} + v_1 \frac{\partial^2}{\partial y^2} + w_1 \frac{\partial^2}{\partial y \partial z} + \kappa^2 v_1 \right) \\
&\qquad\qquad \times \frac{e^{-\iota\kappa(r-Vt)}}{r} d\xi\, d\eta\, d\zeta \\
Z &= - \frac{\iota V}{\kappa} \iiint \left(u_1 \frac{\partial^2}{\partial x \partial z} + v_1 \frac{\partial^2}{\partial y \partial z} + w_1 \frac{\partial^2}{\partial z^2} + \kappa^2 w_1 \right) \\
&\qquad\qquad \times \frac{e^{-\iota\kappa(r-Vt)}}{r} d\xi\, d\eta\, d\zeta
\end{aligned} \right\} \dots(5).$$

The expressions for the components of the magnetic force can be obtained in a similar way, and are

$$\alpha = \iiint \left(w_1 \frac{\partial}{\partial y} - v_1 \frac{\partial}{\partial z} \right) \frac{e^{-\iota\kappa(r-Vt)}}{r} d\xi\, d\eta\, d\zeta, \text{ etc.}$$

These expressions only include the parts of the electric and magnetic forces which depend on the electrical oscillations of period $2\pi/\kappa V$. When there are oscillations of more than one period, the components of convection current u, v, w will be expressible in the form

$$\Sigma u e^{\iota\kappa Vt}, \quad \Sigma v e^{\iota\kappa Vt}, \quad \Sigma w e^{\iota\kappa Vt},$$

and the expressions for the electric and magnetic forces are obtained from the above by summation. The F, G, H of the above investigation is different from Maxwell's vector potential, the difference being a vector whose components are of the form $\frac{\partial \chi}{\partial x}$, $\frac{\partial \chi}{\partial y}$, $\frac{\partial \chi}{\partial z}$. The integrals in the form (5) admit of easy interpretation. From the expressions given in (5) it immediately appears that the components of the electric force at a point are the sums of the components of the electric forces due to a distribution of Hertzian elements throughout the space where there are convection currents, the direction and strength of an element being that of the convection current at the point. The solution in this form is thus the analogue of the solution $V = \iiint \frac{\rho}{r} d\xi \, d\eta \, d\zeta$ in the theory of potential.

12. It is known that even when the vibrations are slow the currents in a conductor tend to concentrate in the neighbourhood of the surface, and, as the vibrations become faster, the currents come to be practically confined to a thin layer at the surface; the limit of this state of affairs is a perfect conductor. In such a conductor the magnetic and electric forces will be zero at each point inside it and at the surface the electric force will be normal to it and the magnetic force tangential to it. The currents are on the surface and are measured by the discontinuity in the magnetic force in crossing the surface; the relation between current strength and magnetic force at the surface is found as follows. Let the axis of z be the normal to the surface at a point on it drawn outwards, the axes of x and y being in the tangent plane to the surface at the point. Then imagining the current to be the limit of a current distributed throughout a small thickness, the relations between current and magnetic force at any point throughout this thickness are given by

$$\frac{1}{V^2} \frac{\partial X}{\partial t} + 4\pi u = \frac{\partial \gamma}{\partial y} - \frac{\partial \beta}{\partial z}, \text{ etc.,}$$

where u is indefinitely great, the thickness being supposed indefinitely small. Integrating throughout the thickness, the first of the above relations becomes

$$\int \frac{1}{V^2} \frac{\partial X}{\partial t}\, dz + 4\pi \int u dz = \int \frac{\partial \gamma}{\partial y}\, dz - \int \frac{\partial \beta}{\partial z}\, dz.$$

The integral $\int \frac{1}{V^2} \frac{\partial X}{\partial t}\, dz$ vanishes in the limit, for the quantity under the sign of integration remains finite while the range of integration ultimately vanishes; similarly the integral $\int \frac{\partial \gamma}{\partial y}\, dz$ vanishes and therefore

$$4\pi \int u dz = -\bar{\beta},$$

where $\bar{\beta}$ is the value of β on the surface, its value at an internal point being zero. In the same way it may be shewn that

$$4\pi \int v dz = \bar{a},$$

where $\bar{a}$ is the value of α on the surface. Therefore the current on the surface is measured by the magnetic force at the surface divided by 4π, their directions being perpendicular and such that the directions of the magnetic force, the current and the outward drawn normal to the surface form a right-handed system of axes. In what follows perfect conductors will be chiefly considered, as in the case of waves whose vibrations are fairly fast the effect of imperfect conduction is a slight dissipation of the energy of vibration into the conductors, which may in general be left out of account.

13. The expressions obtained, § 11, for the electric force will, so far as they depend on conductors, be simplified when the conductors are taken to be perfect. In this case the currents, as was seen above, are zero everywhere throughout the volume of a conductor and exist only on the surface, so that the volume integrals in (5) due to the presence of perfect conductors become surface integrals. Let l_1, m_1, n_1 be the direction cosines of the current at a point ξ, η, ζ on the surface of a conductor and M the magnetic force at the point, then the values of X, Y, Z, the components of the electric force at any point x, y, z of the medium due to the conductors, become

$$
\left.\begin{aligned}
X &= -\frac{\iota V}{4\pi\kappa}\iint M\left(l_1\frac{\partial^2}{\partial x^2} + m_1\frac{\partial^2}{\partial x\partial y} + n_1\frac{\partial^2}{\partial x\partial z} + \kappa^2 l_1\right)\frac{e^{-\iota\kappa r}}{r}\,dS \\
Y &= -\frac{\iota V}{4\pi\kappa}\iint M\left(l_1\frac{\partial^2}{\partial x\partial y} + m_1\frac{\partial^2}{\partial y^2} + n_1\frac{\partial^2}{\partial y\partial z} + \kappa^2 m_1\right)\frac{e^{-\iota\kappa r}}{r}\,dS \\
Z &= -\frac{\iota V}{4\pi\kappa}\iint M\left(l_1\frac{\partial^2}{\partial x\partial z} + m_1\frac{\partial^2}{\partial y\partial z} + n_1\frac{\partial^2}{\partial z^2} + \kappa^2 n_1\right)\frac{e^{-\iota\kappa r}}{r}\,dS
\end{aligned}\right\} \quad (6),
$$

where the integrals are now taken over the surfaces of the conductors. The components of the magnetic force due to the conductors are given by

$$
\left.\begin{aligned}
\alpha &= \frac{1}{4\pi}\iint M\left(n_1\frac{\partial}{\partial y} - m_1\frac{\partial}{\partial z}\right)\frac{e^{-\iota\kappa r}}{r}\,dS \\
\beta &= \frac{1}{4\pi}\iint M\left(l_1\frac{\partial}{\partial z} - n_1\frac{\partial}{\partial x}\right)\frac{e^{-\iota\kappa r}}{r}\,dS \\
\gamma &= \frac{1}{4\pi}\iint M\left(m_1\frac{\partial}{\partial x} - l_1\frac{\partial}{\partial y}\right)\frac{e^{-\iota\kappa r}}{r}\,dS
\end{aligned}\right\} \quad\ldots\ldots\ldots (7).
$$

Thus, when there are perfect conductors only in the medium, the problem of finding the circumstances of propagation of waves of given period through it is reduced to finding the magnetic force at each point of the surface of the conductors.

14. In the preceding investigation of the integrals of the equations everything is expressed in terms of the convection currents which may be looked upon as discontinuities in the time rate of change of the electric force; in exactly the same way integrals of the equations can be found expressing everything in terms of discontinuities in the time rate of change of the magnetic induction. The expressions which correspond to (6) and (7) now make the magnetic force tangential to the surface vanish, while the electric force tangential to it is discontinuous. From this it appears that, if at each point of a surface, which encloses all the sources of the waves, the electric and magnetic forces tangential to it be known, the electric and magnetic forces at every point outside it can be expressed in terms of them, the components of the electric force being given by

$$X = -\frac{\iota V}{4\pi\kappa}\iint M\left(l_1\frac{\partial^2}{\partial x^2} + m_1\frac{\partial^2}{\partial x\partial y} + n_1\frac{\partial^2}{\partial x\partial z} + \kappa^2 l_1\right)\frac{e^{-\iota\kappa r}}{r}\,dS$$

$$+\frac{1}{4\pi}\iint E\left(n_1'\frac{\partial}{\partial y} - m_1'\frac{\partial}{\partial z}\right)\cdot\frac{e^{-\iota\kappa r}}{r}\,dS, \text{ etc.,}$$

where E and M are the electric and magnetic forces tangential to the surface, l_1, m_1, n_1 the direction cosines of the tangent to the surface perpendicular to M, and l_1', m_1', n_1' the direction cosines of the tangent perpendicular to E.

15. The electric and magnetic forces at any point when an arbitrary disturbance is being propagated through the medium can be obtained in a similar way. As in § 11 only those parts of the electric and magnetic forces, which belong to the propagation of the disturbance, will be taken into account. Assume, as in that case,

$$J = -\frac{1}{V^2}\frac{\partial\phi}{\partial t},$$

then the equations to be satisfied by F, G, H are (4'), whence

$$F = \iiint\frac{u_1}{r}\,d\xi\,d\eta\,d\zeta,$$

$$G = \iiint\frac{v_1}{r}\,d\xi\,d\eta\,d\zeta,$$

$$H = \iiint\frac{w_1}{r}\,d\xi\,d\eta\,d\zeta,$$

where u_1, v_1, w_1 are the values of u, v, w at the point ξ, η, ζ at a time r/V before the time t under consideration. Now

$$\frac{\partial\phi}{\partial t} = -V^2\iiint\left(\frac{\partial}{\partial x}\frac{u_1}{r} + \frac{\partial}{\partial y}\frac{v_1}{r} + \frac{\partial}{\partial z}\frac{w_1}{r}\right)d\xi\,d\eta\,d\zeta,$$

and, writing

$$\bar{u}_1 = \int^t u_1\,dt, \quad \bar{v}_1 = \int^t v_1\,dt, \quad \bar{w}_1 = \int^t w_1\,dt,$$

the parts of the components of the electric force which depend on the propagation of the disturbance are given by

$$
\left.
\begin{aligned}
X &= V^2 \iiint \left(\frac{\partial^2}{\partial x^2} \frac{\bar{u}_1}{r} + \frac{\partial^2}{\partial x \partial y} \frac{\bar{v}_1}{r} + \frac{\partial^2}{\partial x \partial z} \frac{\bar{w}_1}{r} - \frac{1}{V^2} \frac{\partial^2}{\partial t^2} \frac{\bar{u}_1}{r} \right) d\xi\, d\eta\, d\zeta \\
Y &= V^2 \iiint \left(\frac{\partial^2}{\partial x \partial y} \frac{\bar{u}_1}{r} + \frac{\partial^2}{\partial y^2} \frac{\bar{v}_1}{r} + \frac{\partial^2}{\partial y \partial z} \frac{\bar{w}_1}{r} - \frac{1}{V^2} \frac{\partial^2}{\partial t^2} \frac{\bar{v}_1}{r} \right) d\xi\, d\eta\, d\zeta \\
Z &= V^2 \iiint \left(\frac{\partial^2}{\partial x \partial z} \frac{\bar{u}_1}{r} + \frac{\partial^2}{\partial y \partial z} \frac{\bar{v}_1}{r} + \frac{\partial^2}{\partial z^2} \frac{\bar{w}_1}{r} - \frac{1}{V^2} \frac{\partial^2}{\partial t^2} \frac{\bar{w}_1}{r} \right) d\xi\, d\eta\, d\zeta
\end{aligned}
\right\} \quad (8).
$$

These expressions can be modified for the case of perfect conductors as in § 13. The part of the electric force at any point which depends on the disturbance is thus expressed in terms of the convection currents which existed at the various points at a time $t - r/V$, where r is the distance of the point under consideration from any other point. The effect of a disturbance at a point A travels out from it in spherical waves, arriving at a point P in the time AP/V, when P takes up the disturbance, after which time it comes to rest*.

16. In the above a single medium filling all space in which there are present what have been termed convection currents is contemplated, and on this hypothesis it has been shewn that, assuming Faraday's laws, electrical effects, whether the propagation of oscillations of definite periods or of an arbitrary disturbance is considered, are propagated in the medium with a definite velocity, the directions of propagation being straight lines and the displacements strictly transverse. Further it has been shewn that the circumstances of the propagation of electrical effects can be completely expressed in terms of the distribution, supposed known for all time, of the convection currents, these convection currents corresponding mathematically to the singularities of the necessary functions. The effect at a point P is the sum of the effects, due to all points Q, which depend on the state of affairs at Q at a previous time

---

* For the complete expressions for the components of electric force when the disturbances are supposed to be due to moving charges, see Appendix C.

$t - PQ/V$; the assumption* is thus implicitly involved that Maxwell's aethereal displacement current is independent of the motion of the aether if there is such a motion. By the application of Faraday's law, the electric force which would act on an element of a circuit at any point of the aether can be obtained, and this is on Maxwell's theory taken to be the electric force at that point in the aether, the element of circuit being conceived of as fixed.

* See § 22.

CHAPTER III.

CONVECTION CURRENTS IN MOTION AND MATERIAL MEDIA.

17. To obtain expressions for the electric force on an element of a convection current moving in a known manner, let an element of such a current at a point x, y, z, the axes of reference being supposed to be fixed in space *, be considered and let l, m, n be the direction cosines of the element. Applying Faraday's law and remembering that it is supposed to hold for closed circuits, the line integral of the electric force in the direction of the element acting on it is

$$- \frac{d}{dt} \int (lF + mG + nH) \frac{ds}{d\sigma} \, d\sigma,$$

where F, G, H define Maxwell's vector-potential or electrokinetic momentum, σ is a coordinate defining the order of the element in the circuit, and where, since the convection current is not supposed to be at rest, l, m, n may vary with the time. The electric force in the direction of the element is therefore given by the expression

$$- l \left(\frac{\partial F}{\partial t} + p \frac{\partial F}{\partial x} + q \frac{\partial F}{\partial y} + r \frac{\partial F}{\partial z} \right) - m \left(\frac{\partial G}{\partial t} + p \frac{\partial G}{\partial x} + q \frac{\partial G}{\partial y} + r \frac{\partial G}{\partial z} \right)$$

$$- n \left(\frac{\partial H}{\partial t} + p \frac{\partial H}{\partial x} + q \frac{\partial H}{\partial y} + r \frac{\partial H}{dz} \right) - F \frac{\partial p}{\partial s} - G \frac{\partial q}{\partial s} - H \frac{\partial r}{\partial s} - \frac{\partial \phi}{\partial s},$$

* Appendix A.

where p, q, r are the component velocities of the element of convection current considered. Using the relations

$$\alpha = \frac{\partial H}{\partial y} - \frac{\partial G}{\partial z},$$

$$\beta = \frac{\partial F}{\partial z} - \frac{\partial H}{\partial x},$$

$$\gamma = \frac{\partial G}{\partial x} - \frac{\partial F}{\partial y},$$

the above expression for the electric force in the direction of the element becomes

$$-l\left(\frac{\partial F}{\partial t} - \gamma q + \beta r\right) - m\left(\frac{\partial G}{\partial t} - \alpha r + \gamma p\right) - n\left(\frac{\partial H}{\partial t} - \beta p + \alpha q\right)$$

$$-p\left(l\frac{\partial F}{\partial x} + m\frac{\partial F}{\partial y} + n\frac{\partial F}{\partial z}\right) - q\left(l\frac{\partial G}{\partial x} + m\frac{\partial G}{\partial y} + n\frac{\partial G}{\partial z}\right)$$

$$-r\left(l\frac{\partial H}{\partial x} + m\frac{\partial H}{\partial y} + n\frac{\partial H}{\partial z}\right) - F\frac{\partial p}{\partial s} - G\frac{\partial q}{\partial s} - H\frac{\partial r}{\partial s} - \frac{\partial \phi}{\partial s},$$

which is equivalent to

$$-l\left(\frac{\partial F}{\partial t} - \gamma q + \beta r\right) - m\left(\frac{\partial G}{\partial t} - \alpha r + \gamma p\right) - n\left(\frac{\partial H}{\partial t} - \beta p + \alpha q\right)$$

$$-\frac{\partial}{\partial s}(Fp + Gq + Hr + \phi).$$

Hence the components of the electric force, which acts on an element of convection current at the point x, y, z moving with a velocity whose components are p, q, r, are given by[*]

---

[*] It will be observed that it is the time rate of variation of the vector F, G, H in a direction, which itself changes with the motion, that is taken, not the time rate of variation in a fixed direction. It is the first of these that is appropriate to electrodynamics, the experimental laws (Faraday's) of which are expressed in terms of the behaviour of closed circuits. Some writers have erroneously used the second; their results however are in general correct, as for the case usually treated of, that of uniform motion, the expressions are the same.

$$\left.\begin{array}{c} -\dfrac{\partial F}{\partial t} + \gamma q - \beta r - \dfrac{\partial \chi}{\partial x} \\[2mm] -\dfrac{\partial G}{\partial t} + \alpha r - \gamma p - \dfrac{\partial \chi}{\partial y} \\[2mm] -\dfrac{\partial H}{\partial t} + \beta p - \alpha q - \dfrac{\partial \chi}{\partial z} \end{array}\right\} \dots \dots \dots \dots (1),$$

where $\qquad \chi = Fp + Gq + Hr + \phi.$

18. On any theory all electrical effects must be explicable on the hypothesis of one uniform medium in which there are convection currents; thus the difference between the aether and any material medium must be the presence of convection currents, that is, wherever there is matter there are these convection currents distributed in the aether, which are to be regarded as discontinuities in it. A complete knowledge of the distribution and strengths of these convection currents for all time would make it possible to determine completely all the electrical circumstances, but such knowledge it is impossible to obtain, and as in any case what can be observed is the effect of an aggregate of these convection currents, it is necessary to make hypotheses which shall replace a knowledge of the individual by something which shall represent the effect of the aggregate. It has been remarked by Maxwell* that "if we attempt to extend our theory to the case of dense media, we become involved not only in all the ordinary difficulties of molecular theories, but in the deeper mystery of the relation of the molecules to the electromagnetic medium." This remark is clearly equally applicable to the case of any material medium; and the assumption made in his next paragraph, it being remembered that throughout this chapter Maxwell is thinking of the medium as at rest except in so far as there are motions due to electrical disturbances, is equivalent to assuming that for transparent media, the aggregate effect of the convection currents can be represented by displacement currents in the

---

* Treatise, Vol. II. § 794.

aether. These displacement currents, which replace the convection currents and which, to distinguish them, may be termed material displacement currents, differ from aethereal displacement currents inasmuch as they must be conceived of as moving with the velocity of the aggregate of the convection currents which they replace, and as being acted on by the electric force which would act on a convection current moving with that velocity. This, combined with the assumption that the relation between the electric force acting on material displacement current and material displacement is of the same kind as that existing between aethereal electric force and aethereal displacement, suffices to determine the circumstances of the propagation of electrical effects through any transparent material medium.

19. Let f, g, h denote the components of the aethereal displacement at the point x, y, z referred to axes fixed in space, f_1, g_1, h_1 the components of the material displacement which replaces the convection currents distributed throughout the medium, and F, G, H the components of the electrokinetic momentum. The components of the aethereal displacement current then are

$$\frac{\partial f}{\partial t}, \quad \frac{\partial g}{\partial t}, \quad \frac{\partial h}{\partial t},$$

and the components of the electric force in the aether are

$$-\frac{\partial F}{\partial t} - \frac{\partial \phi}{\partial x}, \quad -\frac{\partial G}{\partial t} - \frac{\partial \phi}{\partial y}, \quad -\frac{\partial H}{\partial t} - \frac{\partial \phi}{\partial z},$$

giving

$$4\pi V^2 f = -\frac{\partial F}{\partial t} - \frac{\partial \phi}{\partial x}$$
$$4\pi V^2 g = -\frac{\partial G}{\partial t} - \frac{\partial \phi}{\partial y} \left. \right\} \quad \dots\dots\dots\dots (2).$$
$$4\pi V^2 h = -\frac{\partial H}{\partial t} - \frac{\partial \phi}{\partial z}$$

The components of the material displacement current are

$$\frac{df_1}{dt}, \quad \frac{dg_1}{dt}, \quad \frac{dh_1}{dt},$$

where $\dfrac{df_1}{dt}$ denotes the time rate of variation of f_1, the motion being taken into account, and the components of the electric force acting on this material displacement current are by § 17

$$-\frac{\partial F}{\partial t} + \gamma q - \beta r - \frac{\partial \chi}{\partial x},$$

$$-\frac{\partial G}{\partial t} + \alpha r - \gamma p - \frac{\partial \chi}{\partial y},$$

$$-\frac{\partial H}{\partial t} + \beta p - \alpha q - \frac{\partial \chi}{\partial z},$$

whence for an isotropic material medium

$$\left.\begin{aligned}
\frac{4\pi V^2}{\kappa} f_1 &= -\frac{\partial F}{\partial t} + \gamma q - \beta r - \frac{\partial \chi}{\partial x} \\[2mm]
\frac{4\pi V^2}{k} g_1 &= -\frac{\partial G}{\partial t} + \alpha r - \gamma p - \frac{\partial \chi}{\partial y} \\[2mm]
\frac{4\pi V^2}{\kappa} h_1 &= -\frac{\partial H}{\partial t} + \beta p - \alpha q - \frac{\partial \chi}{\partial z}
\end{aligned}\right\} \dots\dots\dots(3).$$

The components of the total current are

$$\frac{\partial f}{\partial t} + \frac{df_1}{dt}, \quad \frac{\partial g}{\partial t} + \frac{dg_1}{dt}, \quad \frac{\partial h}{\partial t} + \frac{dh_1}{dt};$$

whence

$$\left.\begin{aligned}
4\pi\left(\frac{\partial f}{\partial t} + \frac{df_1}{dt}\right) &= \frac{\partial \gamma}{\partial y} - \frac{\partial \beta}{\partial z} \\[2mm]
4\pi\left(\frac{\partial g}{\partial t} + \frac{dg_1}{dt}\right) &= \frac{\partial \alpha}{\partial z} - \frac{\partial \gamma}{\partial x} \\[2mm]
4\pi\left(\frac{\partial h}{\partial t} + \frac{dh_1}{dt}\right) &= \frac{\partial \beta}{\partial x} - \frac{\partial \alpha}{\partial y}
\end{aligned}\right\} \dots\dots\dots\dots (4).$$

In the above it has been assumed that there is no closed convection current at the point considered; when there is, the α, β, γ in equations (4) are related to the F, G, H in the same way as magnetic force is related to vector-potential in the theory of magnetism, while in equations (2) and (3) α, β, γ have to be replaced by a, b, c, where

$$a = \frac{\partial H}{\partial y} - \frac{\partial G}{\partial z}, \qquad b = \frac{\partial F}{\partial z} - \frac{\partial H}{\partial x}, \qquad c = \frac{\partial G}{\partial x} - \frac{\partial F}{\partial y}.$$

The vector whose components are a, b, c is the magnetic force in the aether and includes the effect of the local convection currents, whereas the vector α, β, γ is the magnetic force in the ordinary sense which only takes account of the effect of the aggregate, and, to determine the circumstances, the relation between these two vectors must be known, the usual assumption being that they are connected by the relations

$$a = \mu\alpha, \qquad b = \mu\beta, \qquad c = \mu\gamma,$$

where μ is a constant.

The equations which determine the circumstances of the propagation of waves through an isotropic material medium can now be formed, and it will be sufficient to take the case in which the matter is supposed to be moving parallel to the axis of x with a uniform velocity p. The equations (2) and (3) become

$$4\pi V^2 f = -\frac{\partial F}{\partial t} - \frac{\partial \phi}{\partial x}, \text{ etc.,}$$

$$\frac{4\pi V^2}{\kappa} f_1 = -\frac{\partial F}{\partial t} - p\frac{\partial F}{\partial x} - \frac{\partial \phi}{\partial x}, \text{ etc.,}$$

whence

$$4\pi \left(\frac{\partial f}{\partial t} + \frac{df_1}{dt}\right) = -\frac{1}{V^2}\left[\frac{\partial^2 F}{\partial t^2} + \kappa \left(\frac{\partial}{\partial t} + p\frac{\partial}{\partial x}\right)^2 F\right]$$

$$- \frac{1}{V^2}\left[\frac{\partial}{\partial x}\left(\frac{\partial \phi}{\partial t} + \kappa \frac{\partial \phi}{\partial t} + \kappa p\frac{\partial \phi}{\partial x}\right)\right], \text{ etc.}$$

Again from (4)

$$4\pi \left(\frac{\partial f}{\partial t} + \frac{df_1}{dt}\right) = \frac{\partial \gamma}{\partial y} - \frac{\partial \beta}{\partial z}, \text{ etc.,}$$

that is

$$4\pi \left(\frac{\partial f}{\partial t} + \frac{df_1}{dt}\right) = \frac{1}{\mu}\left[\frac{\partial}{\partial x}\left(\frac{\partial F}{\partial x} + \frac{\partial G}{\partial y} + \frac{\partial H}{\partial z}\right) - \nabla^2 F\right], \text{ etc.;}$$

therefore

$$\frac{1}{\mu}\left[\nabla^2 F - \frac{\partial}{\partial x}\left(\frac{\partial F}{\partial x} + \frac{\partial G}{\partial y} + \frac{\partial H}{\partial z}\right)\right] = \frac{1}{V^2}\left[\frac{\partial^2 F}{\partial t^2} + \kappa\left(\frac{\partial}{\partial t} + p\frac{\partial}{\partial x}\right)^2 F\right]$$
$$+ \frac{1}{V^2}\frac{\partial}{\partial x}\left[(\kappa + 1)\frac{\partial\phi}{\partial t} + \kappa p\frac{\partial\phi}{\partial x}\right],$$

with two similar equations. These can be replaced by equations to determine the magnetic force, giving

$$\frac{1}{\mu}\nabla^2\alpha = \frac{1}{V^2}\left[\frac{\partial^2\alpha}{\partial t^2} + \kappa\left(\frac{\partial}{\partial t} + p\frac{\partial}{\partial x}\right)^2\alpha\right],$$

with two similar equations. For waves propagated in the direction of the axis of x the velocity of propagation V' satisfies the equation

$$\frac{V^2}{\mu} = V'^2 + \kappa(V' - p)^2,$$

agreeing to a first approximation with the formula of Fresnel.

In the free aether $\mu = 1$ and $\kappa = 0$, and the equations to determine the circumstances of the propagation of waves through it are

$$\nabla^2\alpha = \frac{1}{V^2}\frac{\partial^2\alpha}{\partial t^2}, \text{ etc.,}$$

which equations shew that the wave surfaces are spheres; thus to an observer moving through the aether the effect of his motion is the same as if he were supposed to be at rest and the centres of the spheres, which form the wave surfaces, were supposed to move relatively to him with his velocity relatively to the direction in which the rays are travelling through the aether; this is Bradley's law of aberration.

20. The equations (2) and (3) of the preceding will hold for any material medium, whether transparent or not, but the equations (4) will require modification when the effect of the aggregate of the convection currents cannot be replaced by a material displacement current. In this case in addition to the

material displacement current there will be a conduction current which must be added to the displacement currents to give the total current in equations (4), and further the relation between this conduction current and the electric force acting on the convection currents must be known; the scheme of equations (2), (3), (4) thus supplemented will include all cases. They are then the same as those used by Larmor, *Phil. Trans.* (A), 1897, pp. 205—300, and of necessity lead to the same expressions for the mechanical forces acting on material media.

CHAPTER IV.

MAXWELL'S DYNAMICAL THEORY OF ELECTRICITY.

21. AT the end of the fourth chapter of the second volume of his treatise Maxwell, having given an account of the phenomena of the induction of currents and of the laws which they obey, states that he "proposes to examine the consequences of the assumption that the phenomena of the electric current are those of a moving system" and "to deduce the main structure of the theory of electricity from a dynamical hypothesis of this kind." The fifth chapter is taken up with assigning physical meanings to the relations occurring in Lagrange's dynamical method. He then proceeds to discuss in the sixth chapter the form of the kinetic energy function for a system of conducting circuits and shews from experimental considerations that the function is the sum of two homogeneous quadratic functions, one involving all the current strengths and the other all the velocities specifying the motion of the conductors. In the seventh chapter the coefficients in that part of the kinetic energy function, which depends on the current strengths only and which is termed the electrokinetic energy, are identified; the electrokinetic momentum of a circuit is defined and the electric force in and the mechanical force on a circuit are deduced. So far Maxwell has confined himself to the consideration of conducting circuits, but in the eighth chapter he proceeds to discuss what takes place in the space outside these conductors. The field is explored by means of various configurations of the secondary circuit and in this way arises the conception of a vector having a determinate direction and magnitude at each point of space and whose line integral round any circuit is the electrokinetic momentum of that circuit. From this is deduced another vector whose surface integral is equal to the line integral of the former round the edge of the surface and which is identified with the magnetic induction of Faraday. The expressions for the components of the electric force at a point in a conductor moving in any manner and for the mechanical force on the element of the conductor at the point are obtained. Maxwell is careful to point out that " the current in all these cases is to be understood as the actual current which includes not only the current of conduction but the current due to variation of the electric displacement "; to this ought to be added the current due to moving electric charges when these exist. In the ninth chapter these results are recapitulated and he then proceeds, §§ 606, 607, to obtain the mathematical connexion between the electrokinetic momentum and the current strengths, which is equivalent to determining the coefficients in the expression for the electrokinetic energy. The remainder of the chapter is occupied with the discussion of the potential energy and the dissipation functions.

22. The chief difficulty of the theory is what is meant by electric displacement and displacement current; it is then important to inquire whether the theory really determines what they are. The fundamental idea both with Faraday and Maxwell is that of a medium (the aether) filling all space and which is the means by which electrical changes in one place produce electrical changes at another place; this postulates the possibility of electrical effects in the aether. Now the kinetic energy of a system in which there are present electric currents contains no terms which are the products of the current strengths and of the velocities of the circuits; further, on the dynamical theory, the current strengths must be conceived as velocities which are the time rates of variation of coordinates specifying electrical degrees of freedom, these coordinates being dynamically independent of the coordinates which define the positions of the circuits in space. From this it follows that the existence of electrical effects in the aether necessitates the existence of electrical degrees of freedom of the aether, the electrical circumstances at each point of space depending on the coordinates which specify these electrical degrees of freedom. The method of exploring the field by means of a circuit does not determine the electrokinetic momentum of a circuit conceived of as moving with the aether but of a circuit fixed in space; thus the electrical coordinates whose existence was shewn above to be necessary must be conceived of as identified with the points of space not as associated with definite elements of the aether. These electrical coordinates will vary from point to point of space, but from the above it follows that for each one of them the space coordinates must be treated as constant, thus their time rates of variation are given by partial differentiation with respect to the time. The time rates of variation of the electrical coordinates define electric currents in the aether at each point of space and, as these currents are vectors, the electrical coordinates at each point of space are determined by a vector. If f, g, h be the components of this vector at the point x, y, z, the axes of reference being supposed fixed in space, the components of the corresponding current at the point are $\dfrac{\partial f}{\partial t}$, $\dfrac{\partial g}{\partial t}$, $\dfrac{\partial h}{\partial t}$. Maxwell's electric displacement in the aether is thus a vector defining the coordinates which specify the electrical degrees of freedom of the aether at a point of space and is quite distinct from the electric displacement associated with material media; the difference between aethereal electric displacement and material electric displacement has already been emphasised*.

23. In obtaining the electric force at a point in the aether the same considerations apply as above; therefore if F, G, H are the components of the electrokinetic momentum at the point x, y, z, the components of the electric force at the point are

$$-\frac{\partial F}{\partial t} - \frac{\partial \phi}{\partial x}, \quad -\frac{\partial G}{\partial t} - \frac{\partial \phi}{\partial y}, \quad -\frac{\partial H}{\partial t} - \frac{\partial \phi}{\partial z}.$$

* Cf. § 18.

The electrokinetic energy belonging to the aether at the point x, y, z is then as in Maxwell's treatise, § 634,

$$\tfrac{1}{2} \left(F \frac{\partial f}{\partial t} + G \frac{\partial g}{\partial t} + H \frac{\partial h}{\partial t} \right) dx\, dy\, dz.$$

Since $\frac{\partial f}{\partial t}, \frac{\partial g}{\partial t}, \frac{\partial h}{\partial t}$ are the components of the electric current in the aether at the point x, y, z, the electric density at the point is

$$\frac{\partial f}{\partial x} + \frac{\partial g}{\partial y} + \frac{\partial h}{\partial z},$$

which is zero unless there is a discontinuity in the aether at the point. In obtaining the potential energy expressed as a function of f, g, h it is sufficient to find it for the electrostatic case; in this case since charge and potential are linearly related the portion of the potential energy contributed by the aether at the point x, y, z is

$$\tfrac{1}{2} (Xf + Yg + Zh)\, dx\, dy\, dz,$$

where X, Y, Z are the components of the electric force at the point and are proportional to f, g, h. Hence the part of the Lagrangian function belonging to the aether at the point x, y, z is

$$\tfrac{1}{2} \left[F \frac{\partial f}{\partial t} + G \frac{\partial g}{\partial t} + H \frac{\partial h}{\partial t} - Xf - Yg - Zh \right] dx\, dy\, dz,$$

and in this form it is properly expressed in terms of the electrical coordinates and their time rates of variation. It ought to be observed however that this Lagrangian function is not expressed in terms of all the coordinates which specify degrees of freedom of the aether; it is a modified* Lagrangian function expressed in terms of coordinates less in number, the original coordinates having been in part eliminated by the process by which the function has been built up. A knowledge of this Lagrangian function does not then supply sufficient data

* See § 31.

from which to develop the dynamics of the aether; when however in addition the convection currents are completely known,
the data are sufficient to determine the relations of the aether
to material media so far as these relations are electrical.

24. The expression for the Lagrangian function given above
is the one which is natural to Maxwell's dynamical theory as
presented in Chapters V. to IX. of the second volume of his
treatise, but in § 635 he gives the second expression

$$\frac{1}{8\pi} \iiint (a\alpha + b\beta + c\gamma)\, dx\, dy\, dz$$

for the electrokinetic energy, the part contributed by the aether
at the point x, y, z being

$$\frac{1}{8\pi} \iiint (\alpha^2 + \beta^2 + \gamma^2)\, dx\, dy\, dz.$$

In § 636 he assigns a reason for choosing this second expression
as the proper one when he says : " The electrokinetic energy of
the system may therefore be expressed either as an integral to
be taken where there are electric currents, or as an integral to
be taken over every part of the field in which magnetic force
exists.   The first integral, however, is the natural expression of
the theory which supposes the currents to act upon each other
directly at a distance, while the second is appropriate to the
theory which endeavours to explain the action between the
currents by means of some intermediate action in the space
between them."   Now in either case it is a modified Lagrangian function which is obtained and the kinetic energy portion
of such a function is made up of portions contributed from
every place where there are time rates of variation of the
coordinates in terms of which the function is expressed, not of
portions contributed from every place where there is motion;
for example, in the theory of the motion of bodies through a
liquid the modified Lagrangian function which is there used
does not express the kinetic energy as an integral taken
throughout the liquid but in terms of the coordinates which

specify the degrees of freedom of the bodies. Thus the expression of the kinetic energy as a sum to which every place where there is motion contributes is unnecessary; as has been already stated, the electrokinetic energy and the kinetic energy of the aether are different things. Further the transformation by which the second expression for the electrokinetic energy is obtained from the first is in general invalid. Maxwell's procedure § 635 is to write for the currents in the first expression their equivalents in terms of the magnetic force and then integrate by parts; in this way the electrokinetic energy is expressed as a volume integral taken throughout all space and a surface integral taken over the surface of the infinitely distant boundary, this latter being omitted on the ground that at a great distance r from the system the components of the magnetic force are of the order of magnitude r^{-3}; but this is only true for the case when all the currents are steady, which is the case Maxwell appears to be thinking of, and then the transformation is legitimate. When the case is that of the propagation of waves, the components of the magnetic force at a great distance r contain terms of the type $e^{\iota\kappa r}/r$, as do the components of the electrokinetic momentum, and the surface integral is no longer negligible, its value being really indeterminate. The argument which has been sometimes used to justify a transformation of this kind in the case of waves— that if all the sources of the disturbance are at a finite distance from the origin, the surface integral over an infinitely distant boundary cannot influence the state of affairs at a finite distance —neglects the fact that the mathematical treatment of trains of waves postulates an infinite time during which the disturbances have been going on, and that therefore, however remote the boundary may be taken, the disturbances have already produced their effect there[*]. It follows then that

---

[*] It is also clear that, if instead of a train of waves a number of disturbances supposed to have been set up at definite times be considered, the transformation is still invalid as the functions necessary to represent this state of affairs are discontinuous and the integration by parts required for the transformation cannot be effected.

Maxwell's second expression for the electrokinetic energy is inadmissible.

25.   The use of Maxwell's second expression for the electrokinetic energy has led to the assumption that Maxwell's electromagnetic theory and MacCullagh's optical theory are the same, the Lagrangian functions of both theories being then identical. This has further led to the identification (tentatively) of the magnetic induction as the velocity of the aether; from what has been said above it follows that both these assumptions are illegitimate, and that conclusions based on them must be rejected. The identification of the magnetic induction as the velocity of the aether has led to the result that the velocity of propagation of light ought to be altered, though possibly to an insensible extent, by a magnetic field*; the use of the proper form of the Lagrangian function (Maxwell's first form), however, leads to equations to determine the propagation of electrical effects, which are unaltered by the introduction of a magnetic field, so that the velocity of propagation of light is unaltered, agreeing with Lodge's experimental result†. The difficulties concerning permanent magnets due to the identification of Maxwell's theory with MacCullagh's likewise disappear. The conclusion then is that MacCullagh's theory is essentially different from Maxwell's and that Maxwell's theory being in agreement with the phenomena is the one which ought to be retained.

---

* Larmor, *Phil. Trans.* (A), 1894.

† If a_1, β_1, γ_1 are the components of the magnetic force and F_1, G_1, H_1 the components of the vector potential of the imposed magnetic field, F_1, G_1, H_1, a_1, β_1, γ_1 are independent of the time. The total components of the electrokinetic momentum are $F+F_1$, $G+G_1$, $H+H_1$ and of the magnetic force $a+a_1$, $\beta+\beta_1$, $\gamma+\gamma_1$; whence if f, g, h are the components of the electric displacement

$$4\pi \dot{f} = \frac{\partial \gamma}{\partial y} - \frac{\partial \beta}{\partial z}, \quad 4\pi \dot{g} = \frac{\partial a}{\partial z} - \frac{\partial \gamma}{\partial x}, \quad 4\pi \dot{h} = \frac{\partial \beta}{\partial x} - \frac{\partial a}{\partial y},$$

since a_1, β_1, γ_1 are derivable from a potential function and

$$4\pi V^2 f = -\frac{\partial F}{\partial t} - \frac{\partial \phi}{\partial x}, \quad 4\pi V^2 g = -\frac{\partial G}{\partial t} - \frac{\partial \phi}{\partial y}, \quad 4\pi V^2 h = -\frac{\partial H}{\partial t} - \frac{\partial \phi}{\partial z}.$$

Therefore the equations to determine a, β, γ or f, g, h are the same as when there is no imposed magnetic field.

CHAPTER V.

DYNAMICAL THEORY.

26. THE tendency of physical investigations has in general been towards the construction of a dynamical theory which shall give a consistent account of phenomena, the path pursued being to arrive at such a result by inductive methods. The fundamental idea underlying attempts at a theory of this kind is that direct knowledge is confined to a knowledge of motions, the other ideas of dynamics, such as force, being inferences which are useful aids in classifying and explaining phenomena in terms of those phenomena which are more intimately known and over which there is more immediate control. Instead of trying to construct a dynamical theory inductively, another mode of proceeding is possible, to assume that all phenomena are to be explained on the basis of a dynamical theory and to proceed from this deductively.

The starting-point then is that the Lagrangian function is a homogeneous quadratic function of the time rates of variation of the coordinates which specify all the degrees of freedom, the coefficients of the expression being functions of these co-ordinates, and that the time integral of this function taken between any two definite times is stationary for the actual motion. This Lagrangian function is necessarily constant for all time and the principles of the Conservation of Energy and of Least Action are included in this statement. What observation reveals in any case is a certain numb r of degrees of freedom of motion and corresponding motions taking place; knowledge is thus confined to a part only of all the degrees of freedom and the question then arises,—What form does the dynamical theory take to fit in with this limited knowledge? For convenience the discussion will be divided into several cases.

27. The degrees of freedom are divided into two sets, one set being specified by coordinates y which are known, the other set being specified by coordinates x which are unknown. The first case to be discussed is that where the Lagrangian function contains no terms of the type $C\dot{x}\dot{y}$ and the coefficients are functions of the y coordinates only. The Lagrangian function L is given by

$$L = T_x + T_y,$$

where
$$T_x = \tfrac{1}{2}\Sigma A_{11}\dot{x}_1{}^2 + \Sigma A_{12}\dot{x}_1\dot{x}_2,$$

$$T_y = \tfrac{1}{2}\Sigma B_{11}\dot{y}_1{}^2 + \Sigma B_{12}\dot{y}_1\dot{y}_2,$$

and the coefficients A_{11}, A_{12}, B_{11}, B_{12}, etc. are functions of the y coordinates only. By the well-known process of ignoration of coordinates the x coordinates can be eliminated and a modified Lagrangian function L' obtained, which is sufficient to determine the motions so far as the y coordinates are concerned. The coordinates x have to be eliminated from the equations

$$\frac{d}{dt}\left(\frac{\partial L}{\partial \dot{y}}\right) - \frac{\partial L}{\partial y} = 0 \quad\ldots\ldots\ldots\ldots\ldots\ldots(1),$$

$$\frac{d}{dt}\left(\frac{\partial L}{\partial \dot{x}}\right) = 0 \quad\ldots\ldots\ldots\ldots\ldots(2),$$

by means of the relations

$$\frac{\partial L}{\partial \dot{x}} = \xi,$$

where the quantities ξ are constant. Writing

$$L = L' + \Sigma\xi\dot{x},$$

it may be shewn that the first set of equations is replaced by

$$\frac{d}{dt}\left(\frac{\partial L'}{\partial \dot{y}}\right) - \frac{\partial L'}{\partial y} = 0 \dots\dots\dots\dots\dots(3),$$

$$\dot{x} = -\frac{\partial L'}{\partial \xi} \quad\dots\dots\dots\dots\dots(4).$$

Now equations (2) are equivalent to

$$A_{11}\dot{x}_1 + A_{12}\dot{x}_2 + \dots = \xi_1 \dots\dots\dots\dots(2'),$$

etc. ;

hence $\qquad \Sigma\xi\dot{x} = A_{11}\dot{x}_1^2 + 2A_{12}\dot{x}_1\dot{x}_2 + \dots ,$

that is $\qquad\qquad \Sigma\xi\dot{x} = 2T_x,$

and therefore $\qquad\quad L' = T_y - T_x,$

in which T_x is supposed to be expressed in terms of the quantities ξ by means of equations (2') and hence is a function of the y coordinates. Thus the motions so far as they depend on the y coordinates are completely determined by the equations

$$\frac{d}{dt}\left(\frac{\partial L'}{\partial \dot{y}}\right) - \frac{\partial L'}{\partial y} = 0,$$

where $\qquad\qquad L' = T - V,$

T is a homogeneous quadratic function of the velocities $\dot{y}$, being that part of the total energy which is due to the motions corresponding to the degrees of freedom specified by the y coordinates, and V is a function of the y coordinates which is equal to that part of the total energy which is due to the motions corresponding to the degrees of freedom whose co-ordinates have been eliminated. From this follows the possibility of the existence of a class of motions whose complete history can be determined from a knowledge of one function, this function being a modified Lagrangian function. An example of this class of motions is furnished by the mechanics of a conservative system of rigid bodies, the kinetic energy of the system being the function denoted above by T and the potential energy the function denoted by V. The Lagrangian function of the motion is $T - V$ and from the above $T + V$ is constant, being equal to the Lagrangian function which is expressed in terms of all the degrees of freedom. On this view then potential energy is the energy of what may be termed the concealed motions, that is the energy of those motions which correspond to degrees of freedom which are not directly observed.

Another example is that of a rigid body or a number of rigid bodies moving through a liquid, the space occupied by the liquid being simply connected and the motion of the liquid irrotational; in this case attention is confined to the degrees of freedom of the moving bodies. In all such cases certain motions corresponding to degrees of freedom which can be specified by coordinates y are observed, and to determine these motions completely it is only necessary to obtain a knowledge of the modified Lagrangian function which is the difference of two functions, these being what are usually termed the kinetic and potential energy functions of the motion. This knowledge though sufficient to determine the motions depending on the observed degrees of freedom, does not suffice to determine the motions depending on the concealed degrees of freedom, the coordinates corresponding to which have been eliminated. The information which is obtained concerning them, when the motion depending on the observed degrees of freedom is of the character here discussed, is that the coordinates specifying these concealed degrees of freedom enter the original Lagrangian function as velocities only, and there are no terms in it which contain a product of velocities one of which belongs to the observed degrees of freedom and the other to the concealed ones.

28. The second case is that in which the Lagrangian function contains no terms of the type $C\dot{x}\dot{y}$ but both kinds of coordinates occur. The Lagrangian function is in this case given by

$$L = T_x + T_y,$$

where
$$T_x = \tfrac{1}{2}\Sigma A_{11}\dot{x}_1{}^2 + \Sigma A_{12}\dot{x}_1\dot{x}_2,$$

$$T_y = \tfrac{1}{2}\Sigma B_{11}\dot{y}_1{}^2 + \Sigma B_{12}\dot{y}_1\dot{y}_2,$$

and the coefficients A_{11}, A_{12}, B_{11}, B_{12}, etc. are functions of both kinds of coordinates x and y. The equations of motion are

$$\frac{d}{dt}\left(\frac{\partial L}{\partial \dot{y}}\right) - \frac{\partial L}{\partial y} = 0 \quad\dots\dots\dots\dots (1),$$

$$\frac{d}{dt}\left(\frac{\partial L}{\partial \dot{x}}\right) - \frac{\partial L}{\partial x} = 0 \quad\dots\dots\dots\dots (2),$$

and writing
$$L = L' + \Sigma \xi \dot{x},$$

where
$$\xi = \frac{\partial L}{\partial \dot{x}} \quad\dots\dots\dots\dots\dots (3),$$

the equations (1) and (2) are replaced by

$$\frac{d}{dt}\left(\frac{\partial L'}{\partial \dot{y}}\right) - \frac{\partial L'}{\partial y} = 0 \quad\dots\dots\dots\dots\dots(4),$$

$$\dot{x} = -\frac{\partial L'}{\partial \xi}, \quad \dot{\xi} = \frac{\partial L'}{\partial x} \quad\dots\dots\dots\dots(5),$$

where as in § 27
$$L' = T_y - T_x,$$

T_x being expressed as a function of the x and y coordinates and the momenta ξ by means of equations (2). In the motions which belong to this class T_y is the energy due to the time ·rates of variation of the coordinates y which are the observed ones, and T_x ($= V$) is what is termed the potential energy, but, instead of being expressed as a function of the x and y co-ordinates and the momenta ξ, it will appear as a function of the x and y coordinates, the momenta ξ being replaced by their equivalents as functions of the x coordinates. Thus in this case the equations of motion

$$\frac{d}{dt}\left(\frac{\partial L'}{\partial \dot{y}}\right) - \frac{\partial L'}{\partial y} = 0^*,$$

where
$$L' = T - V,$$

* The experimental data in this case do not necessarily give the equations of motion in this form though it is always a possible one.

and is a function of the x and y coordinates and the velocities $\dot{y}$, do not in general suffice to completely determine the motion of the system so far as it depends on the y coordinates. The time rates of variation of the x coordinates expressed as functions of the time must in addition be known; an important case is that where the motion in respect of those of the x coordinates which occur in the coefficients is steady, this steady motion being completely known. A system of linear circuits, in which there are electric currents, moving in a given manner is an example of this class of motions.

29. The third case is that in which the Lagrangian function involves terms of the type $C\dot{x}\dot{y}$ but the coefficients are functions of the y coordinates only. In this case the modified Lagrangian function L', which results when the velocities $\dot{x}$ are eliminated by means of the relations

$$\frac{\partial L}{\partial \dot{x}} = \xi,$$

is no longer of the form $T - V$, where T is a homogeneous quadratic function of the velocities $\dot{y}$ and V is a function of the coordinates y; there are present in addition terms which are linear in the velocities $\dot{y}$. As in § 27 a knowledge of the modified Lagrangian function is sufficient to completely determine the motion so far as it depends on the y coordinates. The motion of a system of rigid bodies to which there are attached a number of gyrostats furnishes an example of this class of motions. A further example is that of a number of solids moving through a liquid, the space occupied by the liquid being multiply connected.

30. The remaining case is that in which the Lagrangian function contains terms of the type $C\dot{x}\dot{y}$ and the coefficients are functions of both kinds of coordinates. In this case the modified Lagrangian function is of the same kind as in § 29, but it involves the x coordinates, so that as in § 28 a knowledge of this function, when, as in cases where it results from

observation, it appears as a function of the x and y coordinates
and the $\dot{y}$ velocities, is not in itself sufficient to completely
determine the motion so far as it depends on the y coordinates.
The time rates of variation of the x coordinates as functions of
the time must in addition be known or relations which are
equivalent.

31. When the number of degrees of freedom is finite the
application of the Lagrangian method presents no difficulties.
When, however, there are an infinite number of degrees of
freedom, some means of identifying the coordinate which belongs
to a particular degree of freedom becomes necessary. For
example, if the coordinates can be arranged in a definite order
forming a numberable aggregate or if each is associated with a
definite point of a straight line, or more strictly with the
element of length of the straight line at the point, this length
being measured from a fixed point on it, the coordinate
specifying each degree of freedom can be identified and the
Lagrangian method can be applied. In the first case the
Lagrangian function has the form of an infinite series and in
the second of a simple integral. In the same way a coordinate
or a finite number of coordinates specifying degrees of freedom
can be associated with each point of a given space; the
Lagrangian function is then a triple integral taken throughout
the given space, the coordinates x, y, z of any point are to be
treated as numbers serving to identify the dynamical coordinates
and, in the formation of the dynamical equation by varying the
Lagrangian function, are to be taken as independent of the
time. The degrees of freedom of an element of a continuous
medium can be specified by means of the coordinates of the
point at which the element is taken and coordinates determining
the change in position, size and shape of the element as it
changes its position*. When the Lagrangian function of the
motion is completely known as a function of the velocities
belonging to all these coordinates, the coordinates will be
determined by the resulting equations as functions of the time

---

* Appendix B.

and of the position of each element at a given time. Now this method of specifying the degrees of freedom of a continuous medium, though convenient when the Lagrangian function is completely known, may be unsuitable when only part of the motions can be observed. When part of the motions which are not directly observed is the motion of the medium in bulk, it is more convenient to conceive of the degrees of freedom as specified by the coordinates of each element and coordinates determining the change in size and shape of the element which at any time occupies a given position, these latter coordinates θ being then associated with the points of space, not with the elements of the medium. If from the coordinates θ certain of them denoted by ϕ can be chosen so that the Lagrangian function has no terms occurring in it which are products of the velocities $\dot{\phi}$ and of any of the other velocities, then by §§ 27, 28 the modified Lagrangian function which results from the elimination of all these latter velocities is of the form $T - V$, where T is a homogeneous quadratic function of the velocities $\dot{\phi}$ and V is a function which does not involve velocities. Now the Faraday-Maxwell theory of the electrical behaviour of the aether leads to a Lagrangian function of the form $T - V$, where

$$T = \frac{1}{2} \iiint \left(F \frac{\partial f}{\partial t} + G \frac{\partial g}{\partial t} + H \frac{\partial h}{\partial t} \right) dx\, dy\, dz$$

is the electrokinetic energy and

$$V = \tfrac{1}{2} \iiint (Xf + Yg + Zh)\, dx\, dy\, dz$$

is the potential energy, f, g, h being the components of a vector which is subject at all points of the free aether to the condition

$$\frac{\partial f}{\partial x} + \frac{\partial g}{\partial y} + \frac{\partial h}{\partial z} = 0,$$

and which defines at each point of space two independent coordinates specifying electrical degrees of freedom. It then follows that the coordinates which specify the degrees of freedom of the aether other than the electrical ones appear in

the original unmodified Lagrangian function as velocities only
and further, that there is no term in it which contains a product
of one of these velocities and a velocity belonging to an
electrical coordinate; the original Lagrangian function cannot
itself be constructed from the data.

32. The difference between free aether and matter being,
according to the views here held, the presence of discontinuities
in the aether, the part of the modified Lagrangian function
which is contributed by the spaces occupied by material media
is not the integral of

$$\frac{1}{2}\left(F\frac{\partial f}{\partial t} + G\frac{\partial g}{\partial t} + H\frac{\partial h}{\partial t} - Xf - Yg - Zh\right),$$

taken throughout the whole space. The places at which there
are discontinuities must be omitted from the range of integration and further, the part contributed to the Lagrangian
function by the motions of these discontinuities must be added
to it. This is the same thing as supposing the above integral
taken throughout the whole space and adding to it a part
depending on the discontinuities, so that the Lagrangian
function now takes the form

$$\frac{1}{2}\iiint\left(F\frac{\partial f}{\partial t} + G\frac{\partial g}{\partial t} + H\frac{\partial h}{\partial t} - Xf - Yg - Zh\right)dx\,dy\,dz,$$

together with a function of the coordinates which specify the
degrees of freedom of all the discontinuities and of the corresponding velocities. A complete knowledge of this latter
function expressed in the form specified cannot be obtained,
and it has therefore to be replaced by something which will
represent the effect of the aggregate as already stated, § 18.
In the case of transparent media observation has shewn that
the function to be added can be replaced by

$$\frac{1}{2}\iiint(F\dot{f}_1 + G\dot{g}_1 + H\dot{h}_1 - Xf_1 - Yg_1 - Zh_1)\,dx\,dy\,dz,$$

in which the coordinates x, y, z of any point now serve to
specify motion of the material medium as well as to identify

the electrical coordinates of the aether. Thus in differentiating f_1, g_1, h_1 with respect to the time, x, y, z are to be treated as varying with the time and the equations of motion are those already written, § 19. It has been shewn that these equations lead to a first approximation to Fresnel's formula for the effect of motion of a transparent medium on the propagation of waves of light through it. The constant κ, which occurs in these equations, depends on the distribution of the discontinuities which constitute the material medium and will therefore be altered by anything which changes their configuration. Now a change in the velocity belonging to any one degree of freedom necessitates change in some at least of the other velocities, thus corresponding to a different velocity of translation of a material medium there will be a different configuration of the discontinuities. To obtain the new configuration in terms of the former one, the two steady states of motion must be compared*; when these steady states are similar the result is that obtained by Larmor, *Phil. Trans.* A, 1897. The alteration depends on the square of the ratio of the velocity of translation to that of radiation and contains the explanation of the Michelson-Morley second order experiment. In the case of any material medium the unknown function depending on the degrees of freedom of the discontinuities must be supposed to be replaced by a function involving a less number of coordinates. These coordinates may be either a number of the original ones or be functions of them; in general after elimination the dynamical equations will be of the type treated of in § 30, the modified Lagrangian function not being expressible as the difference of a kinetic energy function and a potential energy function, though it is to be observed that the coordinates which specify degrees of freedom of the aether other than electrical ones do not enter into these equations. The above result in respect of the comparison of the configurations of the discontinuities for two steady states of motion still holds provided these states are similar, and effects such as the dissipation of electric energy in conductors, absorption, etc., are to be

* Appendix C.

explained by changes in the configuration of the discontinuities.

33. The theory developed above is dynamical and starts from the assumption that the Lagrangian function of the motions taking place is a homogeneous quadratic function of the velocities belonging to all the degrees of freedom. A knowledge of a part only of these motions is obtained from experiment and the Lagrangian function built up from experimental data is the modified function which would result from the elimination of the coordinates belonging to the unobserved motions. This modified function in some cases takes the form of the difference of two functions usually termed the kinetic energy and the potential energy of the motion. The Lagrangian function built up from experimental data which is arrived at in the Faraday-Maxwell theory is of this type, when the coordinates specifying degrees of freedom of the aether other than those which have been termed electrical have been eliminated, and involves, in addition to these electrical coordinates, others specifying the degrees of freedom of all the discontinuities in the aether the presence of which constitutes matter. This modified function, if it could be written down, would suffice to determine the history of all electrical changes whether they are classified as belonging to electricity, magnetism, heat, light or chemistry; a complete knowledge of the motions determined by the coordinates belonging to degrees of freedom of the aether which have been eliminated cannot be obtained from it. The difference between magnetic and non-magnetic material media can be explained by the possible differences in the geometrical form of the discontinuities, it being non-magnetic if the discontinuities leave the space simply-connected, magnetic if otherwise. When the effects under consideration are those in material media what is obtained is a Lagrangian function or dynamical equations involving, instead of coordinates specifying all the degrees of freedom of the discontinuities, a less number, the others having been eliminated, and what has to be aimed at is equations involving the least number of coordinates which will effectively represent the changes taking place. When effects are observed which are not included in this representation the number of coordinates in evidence must be increased, these new effects of necessity belonging to the degrees of freedom the coordinates corresponding to which have previously been eliminated. If an effect were observed which could not be assigned to possible motions corresponding to the degrees of freedom of the discontinuities, it would furnish information about the motions of the aether, the coordinates corresponding to which have been eliminated, and hence aid in the explanation of gravitation, which is an effect belonging to these concealed motions.

34. It appears from the above that on a dynamical theory the existence of a potential energy function necessitates the existence of concealed motions the coordinates corresponding to which have been eliminated. Hence on the fixed aether theory of Lorentz, where the aether is supposed to possess only electrical degrees of freedom, there could be no potential energy function contributed by the free aether, whether the ions be supposed to possess inertia or not. Thus the theory is not a dynamical theory. Even if the aether be supposed to have, in addition to the electrical motions, other motions which in comparison with them are negligible, it is difficult to see how these could give rise to a potential energy function contributed by the free aether which would not be negligible in comparison with the kinetic energy function. A further objection to either form of the theory is that no adequate provision is made for the explanation of gravitation. On the other hand, the theory here developed is consistent with the obtaining of equations to determine the electrical motions and of expressions for the mechanical forces due to these motions acting on material media which are the same as those which have been made the basis of his investigations by Larmor, *Phil. Trans.* A, 1897, "Aether and Matter," 1900, and the results there arrived at constitute powerful evidence for the truth of these equations and expressions within their range of applicability. Further, the theory provides for the explanation of gravitation and of other unexplained effects, if there should be any others which are not electrical, by means of the concealed motions of the aether which on this theory give rise to the potential energy function contributed by the free aether. It also shews that, in the investigation of any particular class of effects, attention can be concentrated on the motions accompanying these effects, other motions being treated as belonging to coordinates which have been eliminated, the simplest case being that treated of § 27 where the characteristics are those of a conservative system and which includes the greater number of physical phenomena so far investigated.

CHAPTER VI.

PROPAGATION OF ELECTRICAL EFFECTS IN SIMPLY-CONNECTED SPACES.

35. IN the investigations of this chapter the conductors will be assumed to be perfect conductors, the effect of imperfect conduction in the case of waves of a frequency as high as that of Hertzian waves being negligible. By what has been shewn, § 12, since the tangential electric force at the surface of such a conductor vanishes, the wave fronts must everywhere at the surface of a conductor either be tangential to it or cut it at right angles. Taking first the case where the wave front at the surface of a conductor is tangential to it, it appears that the waves are reflected at the surface of a conductor and then travel outwards from it; thus in order that the oscillations should be permanent the waves must all be reflected again or must be maintained by some external influence, in which case they are not free oscillations. In the case of a single conductor whose exterior surface is everywhere convex a disturbance of this character arriving at the surface is reflected there, and then travels outwards into infinite space so that there can be no permanent free oscillations of this kind belonging to such a conductor. When the exterior surface of the conductor is not everywhere convex or when there are a number of conductors, a portion of the disturbance will be reflected a second time and so on, but the whole disturbance is not reflected; thus there are no permanent free oscillations of the character considered belonging to the space exterior to a system of conductors.

When the wave fronts cut the surface of a conductor at right angles, let a definite wave front be selected and followed in its course; at any instant of time it intersects the surface of a conductor in a closed curve and, as the space is simply connected, this curve reduces itself to a point somewhere on the surface of the conductor, the wave front as this point is approached becomes conical in form and unless at the same time it intersects another conductor the wave will travel outwards into infinite space and the oscillations will not be permanent. Thus for permanent free oscillations the space determined by the conducting surfaces must, if simply connected, be closed and finite in extent. This result might be inferred from the fact that it is the aether which possesses what corresponds to inertia, and that the effect of indefinitely increasing a simply-connected space which is occupied by something which possesses inertia is to make it possible to have waves of all periods propagated through it.

36. In illustration of the foregoing the possible free oscillations in the space between two concentric spherical surfaces will be considered. If ξ, η, ζ denote orthogonal curvilinear coordinates, the equations expressing current strength in terms of magnetic force and magnetic induction in terms of electric force are, § 9,

$$4\pi \iint \left(\frac{u}{h_2 h_3} \, d\eta \, d\zeta + \frac{v}{h_3 h_1} \, d\zeta \, d\xi + \frac{w}{h_1 h_2} \, d\xi \, d\eta \right)$$

$$= \int \left(\frac{\alpha}{h_1} \, d\xi + \frac{\beta}{h_2} \, d\eta + \frac{\gamma}{h_3} \, d\zeta \right),$$

and $$- \frac{d}{dt} \iint \left(\frac{a}{h_2 h_3} \, d\eta \, d\zeta + \frac{b}{h_3 h_1} \, d\zeta \, d\xi + \frac{c}{h_1 h_2} \, d\xi \, d\eta \right)$$

$$= \int \left(\frac{X}{h_1} \, d\xi + \frac{Y}{h_2} \, d\eta + \frac{Z}{h_3} \, d\zeta \right),$$

where u, v, w are the components of current strength, α, β, γ the components of the magnetic force, a, b, c of the magnetic induction and X, Y, Z of the electric force, these components being estimated along the normals to the surfaces defined respectively by $\xi = \text{const.}$, $\eta = \text{const.}$, $\zeta = \text{const.}$, the elements of length in these directions being $\dfrac{d\xi}{h_1}, \dfrac{d\eta}{h_2}, \dfrac{d\zeta}{h_3}$. These are equivalent to

$$
\left.
\begin{aligned}
4\pi u &= h_2 h_3 \left\{ \frac{\partial}{\partial \eta}\left(\frac{\gamma}{h_3}\right) - \frac{\partial}{\partial \zeta}\left(\frac{\beta}{h_2}\right) \right\} \\
4\pi v &= h_3 h_1 \left\{ \frac{\partial}{\partial \zeta}\left(\frac{\alpha}{h_1}\right) - \frac{\partial}{\partial \xi}\left(\frac{\gamma}{h_3}\right) \right\} \\
4\pi w &= h_1 h_2 \left\{ \frac{\partial}{\partial \xi}\left(\frac{\beta}{h_2}\right) - \frac{\partial}{\partial \eta}\left(\frac{\alpha}{h_1}\right) \right\}
\end{aligned}
\right\} \ldots\ldots\ldots (1),
$$

and

$$
\left.
\begin{aligned}
-\frac{\partial a}{\partial t} &= h_2 h_3 \left\{ \frac{\partial}{\partial \eta}\left(\frac{Z}{h_3}\right) - \frac{\partial}{\partial \zeta}\left(\frac{Y}{h_2}\right) \right\} \\
-\frac{\partial b}{\partial t} &= h_3 h_1 \left\{ \frac{\partial}{\partial \zeta}\left(\frac{X}{h_1}\right) - \frac{\partial}{\partial \xi}\left(\frac{Z}{h_3}\right) \right\} \\
-\frac{\partial c}{\partial t} &= h_1 h_2 \left\{ \frac{\partial}{\partial \xi}\left(\frac{Y}{h_2}\right) - \frac{\partial}{\partial \eta}\left(\frac{X}{h_1}\right) \right\}
\end{aligned}
\right\} \ldots\ldots\ldots (2).
$$

In the case of polar coordinates r, θ, ϕ they become

$$
\left.
\begin{aligned}
4\pi u &= \frac{1}{r^2 \sin\theta} \left\{ \frac{\partial}{\partial\theta}(\gamma r \sin\theta) - \frac{\partial}{\partial\phi}(\beta r) \right\} \\
4\pi v &= \frac{1}{r \sin\theta} \left\{ \frac{\partial\alpha}{\partial\phi} - \frac{\partial}{\partial r}(\gamma r \sin\theta) \right\} \\
4\pi w &= \frac{1}{r} \left\{ \frac{\partial}{\partial r}(\beta r) - \frac{\partial\alpha}{\partial\theta} \right\}
\end{aligned}
\right\} \ldots\ldots (3),
$$

and

$$
\left.
\begin{aligned}
-\frac{\partial a}{\partial t} &= \frac{1}{r^2 \sin\theta} \left\{ \frac{\partial}{\partial\theta}(Z r \sin\theta) - \frac{\partial}{\partial\phi}(Y r) \right\} \\
-\frac{\partial b}{\partial t} &= \frac{1}{r \sin\theta} \left\{ \frac{\partial X}{\partial\phi} - \frac{\partial}{\partial r}(Z r \sin\theta) \right\} \\
-\frac{\partial c}{\partial t} &= \frac{1}{r} \left\{ \frac{\partial}{\partial r}(Y r) - \frac{\partial X}{\partial\theta} \right\}
\end{aligned}
\right\} \ldots\ldots (4).
$$

The equations are now in a form which is convenient for the treatment of the space between two concentric spherical surfaces. In the cases to be here treated the concentric spherical surfaces bounding the space will be supposed to be perfectly conducting surfaces and the medium occupying the space will be supposed to be non-magnetic. The magnetic force and the magnetic induction are then identical and the current strength is given in terms of the electric force by the relations

$$u = \frac{1}{4\pi V^2} \frac{\partial X}{\partial t}, \quad v = \frac{1}{4\pi V^2} \frac{\partial Y}{\partial t}, \quad w = \frac{1}{4\pi V^2} \frac{\partial Z}{\partial t},$$

where V is the velocity of propagation of electrical effects through the medium. The conditions to be satisfied at a bounding surface are

$$Y = 0, \quad Z = 0.$$

It is not difficult to see that any case can be built up from the following two simple cases. First $X = 0$, $Y = 0$ throughout the space and on the boundaries, secondly $\alpha = 0$, $\beta = 0$ throughout the space and on the boundaries; these correspond respectively to the typical cases in which the wave fronts are first tangential to the bounding surfaces, secondly, orthogonal to them. Taking the case where only the Z component of the electric force does not vanish, the equations (3) and (4) give

$$\frac{1}{V^2} \frac{\partial Z}{\partial t} = \frac{1}{r} \left\{ \frac{\partial}{\partial r} (\beta r) - \frac{\partial \alpha}{\partial \theta} \right\},$$

$$-\frac{\partial \alpha}{\partial t} = \frac{1}{r^2 \sin \theta} \frac{\partial}{\partial \theta} (Zr \sin \theta),$$

$$-\frac{\partial \beta}{\partial t} = -\frac{1}{r \sin \theta} \frac{\partial}{\partial r} (Zr \sin \theta),$$

$$-\frac{\partial \gamma}{\partial t} = 0.$$

Hence

$$\frac{1}{V^2} \frac{\partial^2 Z}{\partial t^2} = \frac{1}{r \sin \theta} \frac{\partial^2}{\partial r^2} (Zr \sin \theta) + \frac{1}{r^3} \frac{\partial}{\partial \theta} \left\{ \frac{1}{\sin \theta} \frac{\partial}{\partial \theta} (Zr \sin \theta) \right\},$$

that is, writing μ for $\cos \theta$,

$$\frac{1}{V^2} \frac{\partial^2}{\partial t^2} (Zr \sin \theta) = \left(\frac{\partial^2}{\partial r^2} + \frac{1 - \mu^2}{r^2} \frac{\partial^2}{\partial \mu^2} \right) (Zr \sin \theta).$$

Assuming $Zr \sin \theta = \chi e^{\frac{2\pi\iota t}{T}}$ and putting $\frac{2\pi}{VT} = \kappa$, χ satisfies the equation

$$\left(\frac{\partial^2}{\partial r^2} + \frac{1-\mu^2}{r^2}\frac{\partial^2}{\partial\mu^2} + \kappa^2\right)\chi = 0.$$

Therefore

$$\chi = \Sigma R_n (1-\mu^2)\frac{\partial P_n}{\partial\mu},$$

where $P_n(\mu)$ is the zonal harmonic of order n, and n is an integer, because χ has to be finite for all values of μ from -1 to 1 inclusive. The function R_n is given by the equation

$$\frac{\partial^2 R_n}{\partial r^2} + \left(\kappa^2 - \frac{n(n+1)}{r^2}\right)R_n = 0;$$

that is $R_n = r^{\frac{1}{2}}\{AJ_{n+\frac{1}{2}}(\kappa r) + BJ_{-n-\frac{1}{2}}(\kappa r)\}.$

If $r = r_0$ and $r = r_1$ define the bounding surfaces the boundary conditions give

$$AJ_{n+\frac{1}{2}}(\kappa r_0) + BJ_{-n-\frac{1}{2}}(\kappa r_0) = 0,$$

$$AJ_{n+\frac{1}{2}}(\kappa r_1) + BJ_{-n-\frac{1}{2}}(\kappa r_1) = 0,$$

whence $J_{n+\frac{1}{2}}(\kappa r_0)J_{-n-\frac{1}{2}}(\kappa r_1) = J_{n+\frac{1}{2}}(\kappa r_1)J_{-n-\frac{1}{2}}(\kappa r_0),$

an equation to determine κ and thence the possible free periods. When r_0 and r_1 are both finite, this equation has an infinite number of real roots and no others; an interesting particular case is that when $r_0 = 0$, the equation then is

$$J_{n+\frac{1}{2}}(\kappa r_1) = 0,$$

and gives the possible free periods of electrical oscillations in a spherical space. When r_1 becomes indefinitely great $J_{n+\frac{1}{2}}(\kappa r_1)$ tends to the value $\sqrt{\dfrac{2}{\pi\kappa r_1}}\cos\left(\kappa r_1 - \dfrac{n+1}{2}\pi\right)$ and $J_{-n-\frac{1}{2}}(\kappa r_1)$ to the value $\sqrt{\dfrac{2}{\pi\kappa r_1}}\cos\left(\kappa r_1 + \dfrac{n}{2}\pi\right)$, so that the equation to determine the free periods becomes

$$J_{n+\frac{1}{2}}(\kappa r_0)\cos\left(\kappa r_1 + \frac{n\pi}{2}\right) = J_{-n-\frac{1}{2}}(\kappa r_0)\cos\left(\kappa r_1 - \frac{n+1}{2}\pi\right),$$

an equation which is satisfied by all values of κ when r_1 becomes infinite. The interpretation of this is that belonging to the space exterior to a sphere there are no free periods of electrical oscillations of this type, but that waves of any period can be propagated in it.

Taking now the case $\alpha = 0$, $\beta = 0$, where the wave fronts are orthogonal to the bounding surfaces, the equations become

$$4\pi u = \frac{1}{r^2 \sin\theta} \frac{\partial}{\partial\theta} (\gamma r \sin\theta),$$

$$4\pi v = -\frac{1}{r \sin\theta} \frac{\partial}{\partial r} (\gamma r \sin\theta),$$

$$w = 0,$$

$$-\frac{\partial\gamma}{\partial t} = \frac{1}{r} \left\{ \frac{\partial}{\partial r} (Yr) - \frac{\partial X}{\partial\theta} \right\},$$

and it is clear that γ satisfies the same equation as Z does above, hence

$$\gamma r \sin\theta = \Sigma R_n (1 - \mu^2) \frac{\partial P_n}{\partial\mu},$$

where $R_n = r^{\frac{1}{2}} [A J_{n+\frac{1}{2}}(\kappa r) + B J_{-n-\frac{1}{2}}(\kappa r)],$

the boundary conditions now being

$$\frac{\partial}{\partial r} (\gamma r \sin\theta) = 0,$$

when $r = r_0$ and $r = r_1$. Therefore corresponding to the harmonic of order n the equation to determine κ and thence the periods is

$$\frac{\partial}{\partial r_0} \{ r_0^{\frac{1}{2}} J_{n+\frac{1}{2}}(\kappa r_0) \} \frac{\partial}{\partial r_1} \{ r_1^{\frac{1}{2}} J_{-n-\frac{1}{2}}(\kappa r_1) \}$$
$$= \frac{\partial}{\partial r_0} \{ r_0^{\frac{1}{2}} J_{-n-\frac{1}{2}}(\kappa r_0) \} \frac{\partial}{\partial r_1} \{ r_1^{\frac{1}{2}} J_{n+\frac{1}{2}}(\kappa r_1) \},$$

an equation possessing, when r_0 and r_1 are both finite, an infinite number of real roots and no others. When r_0 vanishes the equation is

$$\frac{\partial}{\partial r_1} \{r_1^{\frac{1}{2}} J_{n+\frac{1}{2}} (\kappa r_1)\} = 0,$$

giving the free periods belonging to a spherical space for this type of wave. As before, when r_1 becomes indefinitely great the equation is ultimately satisfied by all values of κ. Thus in the space exterior to a sphere there can be no permanent free electrical oscillations.

37. Corresponding to any closed surface there are, as was seen in § 35, definite free periods belonging to the possible free electrical oscillations which can permanently exist in the space inside. Condensers are a particular case of this, and when the two faces of the condenser are very close together the free periods can in a great number of cases be simply determined. The results for various forms of thin condensers have been given by Larmor[*]. Taking equations (1) and (2) § 36 and choosing the surfaces ζ so that $\zeta = \zeta_0$, $\zeta = \zeta_1$ are the two bounding surfaces of the condenser, a choice which can always be made when the electrostatic problem is solved, the boundary conditions are $X = 0$, $Y = 0$ at both surfaces.

Two cases arise, the one being that for which X and Y both vanish throughout the condenser and for which the wave fronts cut the surfaces of the condenser orthogonally, the other being the conjugate case for which α and β vanish throughout and for which the wave fronts are tangential to the bounding surfaces. In the second case it is clear that the wave lengths will be very short and therefore the periods will be very high, so that the first is the important case; for it equations (1) and (2) become

$$4\pi w = h_1 h_2 \left\{ \frac{\partial}{\partial \xi} \left(\frac{\beta}{h_2} \right) - \frac{\partial}{\partial \eta} \left(\frac{\alpha}{h_1} \right) \right\},$$

$$-\frac{\partial \alpha}{\partial t} = h_2 h_3 \frac{\partial}{\partial \eta} \left(\frac{Z}{h_3} \right),$$

$$-\frac{\partial \beta}{\partial t} = -h_3 h_1 \frac{\partial}{\partial \xi} \left(\frac{Z}{h_3} \right),$$

[*] *Proc. Lond. Math. Soc.* Vol. xxvi. 1894.

whence

$$\frac{1}{V^2} \frac{\partial^2 Z}{\partial t^2} = h_1 h_2 \left[\frac{\partial}{\partial \xi} \left\{ \frac{h_3 h_1}{h_2} \frac{\partial}{\partial \xi} \left(\frac{Z}{h_3} \right) \right\} + \frac{\partial}{\partial \eta} \left\{ \frac{h_2 h_3}{h_1} \frac{\partial}{\partial \eta} \left(\frac{Z}{h_3} \right) \right\} \right].$$

When the surfaces determined by $\zeta = \zeta_0$ and $\zeta = \zeta_1$ are closed surfaces, this equation suffices to determine completely the circumstances, the condition that Z should be everywhere in the medium finite determining the possible free periods. An example is furnished by the case of a spherical condenser; the result in this case is at once deduced from the frequency equation given in § 36 which becomes, when $r_1 - r_0$ is very small,

$$\kappa^2 r_0^2 = n(n+1),$$

n being an integer.

When the bounding surfaces are not closed surfaces, the contours bounding the surfaces may be supposed to be joined by a perfectly conducting surface determined by some relation between ξ and η, the tangential electric force over this surface would have to vanish and this condition would give an equation to determine the free periods. In actual condensers the medium between the two faces is in communication with the medium outside, the ends not being closed; if waves proper to the space closed as above supposed to be set up and then left to themselves, their energy would be rapidly radiated into space, as the conditions imposed suppose them to be reflected at the ends. If the faces at the ends be supposed to be so close together that their distance apart at any point is small compared with the wave length of a possible oscillation, an approximation to the free periods of these possible oscillations can be obtained on the assumption that the ends of the condenser are loops, the condition then being that the magnetic force tangential to the edge of a face vanishes.

When the condenser is formed by two nearly closed surfaces the free periods so found will differ but slightly from those found on the assumption that the surfaces are closed. For example, in the case of the condenser formed by two concentric spheres pierced by small apertures, one in each sphere opposite each other, the frequency equation is

$$n\,(n+1) = \kappa^2 r_0^2,$$

where n is not now an integer but is determined by the condition that

$$\frac{\partial P_n}{\partial \mu} = 0$$

at the edge of the aperture. If α be the angle subtended by the radius of the aperture, supposed circular and very small, at the centre of the sphere, the roots of this equation are given by*

$$n = 1 + k + \frac{\Pi\,(2+k)}{\Pi\,(k)}\,\tan^2\frac{\alpha}{2},$$

where k has all positive integral values including zero and these values differ but little from the integral values of n. It may be inferred by reasoning similar to that of § 35 that in the case of a partially enclosed space, the boundary being supposed to be perfectly conducting, free oscillations with a certain degree of persistence are possible, the persistence being considerable if the linear dimensions of the aperture are small compared with the wave length, or if the aperture is a slit whose width is everywhere small compared with the wave length. An approximation can be obtained in the case of a condenser whose faces are parallel plates at a distance d apart small compared with the wave length λ. The correction for the open end is approximately α which is given by†

$$\alpha = \frac{d}{\pi}\left[\log\frac{\lambda}{\pi d} + \Pi'\,(0) + \tfrac{3}{2}\right],$$

and the magnetic induction in the space between the plates is approximately given by

$$c = A\left[\sin\frac{2\pi}{\lambda}(x-\alpha)\cos nt - \frac{\pi d}{\lambda}\cos\frac{2\pi x}{\lambda}\sin nt\right],$$

---

* Macdonald, *Proc. Lond. Math. Soc.* Vol. xxxi. 1899.
† Appendix D.

where $2\pi/n$ is the period; whence the radiation from the open ends can be calculated.

38. If the two bounding surfaces of the condenser are not closed surfaces, the free periods will be practically unaltered by joining the opposite faces by a very thin wire, whether the ends be supposed closed by a perfectly conducting surface or not. For example, take the case of the condenser formed by two circular plates of equal area, the one being exactly opposite to the other. If r_1 be the radius of either plate, r_0 the radius of a thin wire joining them at their centres, the frequency equation, when the ends are supposed closed by a perfectly conducting

surface, is $$J_n(\kappa r_1) = \frac{J_n(\kappa r_0)}{Y_n(\kappa r_0)} Y_n(\kappa r_1),$$

where n is any integer, and when the ends are not closed is

$$J_n'(\kappa r_1) = \frac{J_n(\kappa r_0)}{Y_n(\kappa r_0)} Y_n'(\kappa r_1).$$

In either case, r_0 being very small, no new free period is introduced and the alteration in those already existing is very small.

When the bounding surfaces of the condenser are closed surfaces a new period is introduced by joining the faces, this period being very long. In the case of the condenser formed by two concentric spherical surfaces, let the wire, supposed to be very thin, be taken as the axis of the harmonics, then for oscillations in which the nodal lines are circles of latitude, the frequency equation is

$$n(n+1) = \kappa^2 r_0^2,$$

where n satisfies the equation

$$P_n(-\cos \alpha) = 0,$$

α being the angle subtended at the centre of the sphere by the radius of the wire. The roots of this equation are given by[*]

$$n = k + \frac{1}{2 \log \dfrac{2}{\alpha}},$$

* Macdonald, *loc. cit.*

where k has all integral values including zero; and corresponding

to
$$n = \frac{1}{2 \log \frac{2}{\alpha}}$$

there is a possible oscillation of very long period.

39. When a constraint is suddenly set up or removed at any point of the medium, the corresponding disturbance is propagated outwards from this point in all directions with the velocity of radiation, each point taking it up as it arrives there and reverting to a steady state after it has passed over it. The complete solution in any case where the original disturbance is confined to a finite space is to be found by the application of the results of § 15; examples have been worked out in detail by Heaviside*. The distinctive feature of such cases is that the disturbance produced at any point lasts for a definite time, which can be very simply determined. When, however, the space in which the disturbance originates is not finite there is no such definite time. An example of this kind is furnished by the suppression of a field of electric force external to a conducting surface, the space outside the conductor being infinitely extended. In this case a continuous series of disturbances arrive at the surface of the conductor, are reflected there and then propagated outwards into space. A knowledge of the reflected disturbances is sufficient to determine the distribution on the surface at any time subsequent to the suppression of the field of force. An example of this kind is that where a uniform field of electric force external to a sphere is suppressed; the expression for the magnetic force due to the reflected disturbances is that given by J. J. Thomson†

$$\gamma = \frac{a \sin \theta}{r} \left\{ 1 - \frac{a}{r} + \frac{a^2}{r^2} \right\} e^{-\frac{(Vt-r)}{2a}} \cos(\phi + \delta),$$

---

* *Electrical Papers*, Vol. II. pp. 375—467.
† *Proc. Lond. Math. Soc.* Vol. XV. p. 210, 1884; *Recent Researches in Electricity and Magnetism*, p. 370.

where
$$\phi = \frac{\sqrt{3}}{2a}(Vt - r),$$

$$\tan \delta = \frac{r - a}{r + a} \tan \frac{\pi}{3};$$

and this expression exists up to a distance from the centre of the sphere $a + Vt$, which defines the greatest distance to which the reflected disturbances have attained at a time t subsequent to the suppression of the field of force. The electrical distribution on the surface of the sphere at any instant immediately follows. The example of an ellipsoidal conductor under the same circumstances has, more particularly for the case when the ellipsoid is of revolution about its greatest axis, been investigated by Abraham*.

40. It appears from § 37 that, when there are condensers whose bounding surfaces are not closed or partially enclosed spaces, the bounding surfaces being in both cases supposed to be perfectly conducting, there are free oscillations belonging to these spaces, which after having been set up have their energy radiated out from the ends of the condenser or through the aperture of the partially enclosed space, the rate at which the oscillations die away depending on the nature of the aperture. The possibility of successive reflexions of the waves in the simply-connected space is the essential condition for the continued existence of free oscillations whether permanent or not; it therefore follows that there will be possible free oscillations, though not permanent, when there is a space occupied by a dielectric medium which is different from the medium surrounding it.

As an illustration let there be two dielectric media, one of them occupying the space inside the surface of a sphere of radius r_0, the other the space outside this surface, and let the ratio of the squares of the velocities of radiation in the two

* *Annalen der Physik und Chemie*, Bd. 66, 1898.

media be K. Then as in § 36 the solution proper to the space inside the sphere, omitting the time factor, is given by

$$\gamma r \sin \theta = A r^{\frac{1}{2}} J_{n+\frac{1}{2}} (\kappa r)(1 - \mu^2) \frac{\partial P_n}{\partial \mu},$$

that case being considered where the wave fronts are orthogonal to the surface of the sphere and where

$$\kappa^2 = \frac{4\pi^2}{V^2 T^2},$$

V being the velocity of radiation in the medium inside the sphere. The solution for the space outside the sphere is given

by $\gamma r \sin \theta = A' r^{\frac{1}{2}} \{J_{-n-\frac{1}{2}} (\lambda r) - e^{-(n+\frac{1}{2}) \pi \iota} J_{n+\frac{1}{2}} (\lambda r)\} (1-\mu^2) \dfrac{\partial P_n}{\partial \mu},$

where $\lambda^2 = \dfrac{4\pi^2}{V'^2 T^2} = K \kappa^2.$

Both media being assumed to be non-magnetic, the boundary conditions are

$$A J_{n+\frac{1}{2}} (\kappa r_0) = A' \{J_{-n-\frac{1}{2}} (\lambda r_0) - e^{-(n+\frac{1}{2}) \pi \iota} J_{n+\frac{1}{2}} (\lambda r_0)\},$$

$$A \frac{\partial}{\partial r_0} \{r_0^{\frac{1}{2}} J_{n+\frac{1}{2}} (\kappa r_0)\} = A' \frac{\partial}{\partial r_0} [r_0^{\frac{1}{2}} \{J_{-n-\frac{1}{2}} (\lambda r_0) - e^{-(n+\frac{1}{2}) \pi \iota} J_{n+\frac{1}{2}} (\lambda r_0)\}],$$

whence the equation to determine κ is

$$\{J_{-n-\frac{1}{2}} (\lambda r_0) - e^{-(n+\frac{1}{2}) \pi \iota} J_{n+\frac{1}{2}} (\lambda r_0)\} \frac{\partial}{\partial r_0} \{r_0^{\frac{1}{2}} J_{n+\frac{1}{2}} (\kappa r_0)\}$$

$$= \frac{\partial}{\partial r_0} [r_0^{\frac{1}{2}} \{J_{-n-\frac{1}{2}} (\lambda r_0) - e^{-(n+\frac{1}{2}) \pi \iota} J_{n+\frac{1}{2}} (\lambda r_0)\}] J_{n+\frac{1}{2}} (\kappa r_0),$$

that is after reduction

$$\kappa J_{n-\frac{1}{2}} (\kappa r_0) \{J_{-n-\frac{1}{2}} (\lambda r_0) - e^{-(n+\frac{1}{2}) \pi \iota} J_{n+\frac{1}{2}} (\lambda r_0)\}$$

$$= - \lambda J_{n+\frac{1}{2}} (\kappa r_0) \{J_{-n+\frac{1}{2}} (\lambda r_0) + e^{-(n+\frac{1}{2}) \pi \iota} J_{n-\frac{1}{2}} (\lambda r_0)\}.$$

It will be sufficient as an illustration to consider the case when $n = 1$; in this case the equation becomes

$$\kappa \sin \kappa r_0 \left(\iota - \frac{1}{\lambda r_0}\right) + \lambda \left(\frac{\sin \kappa r_0}{\kappa r_0} - \cos \kappa r_0\right) = 0,$$

that is $\tan \kappa r_0 = \dfrac{\lambda}{\iota \kappa - \dfrac{\kappa}{\lambda r_0} + \dfrac{\lambda}{\kappa r_0}},$

or
$$\tan \kappa r_0 = \frac{\kappa \sqrt{K}}{\kappa \iota + \dfrac{K-1}{r_0 \sqrt{K}}} \cdot$$

The roots of this equation have their imaginary parts negative, these being very great when K differs but little from unity; thus where the media have their velocities of radiation nearly equal free oscillations die away almost instantaneously, when the velocities of radiation differ considerably the oscillations will persist for some time.

41. The result of the preceding is that, when a space bounded by conductors is simply connected, there are free oscillations belonging to finite spaces occupied by dielectric media, for any given finite space there being definite free periods which in general will be infinite in number. When there are spaces, occupied by dielectric media, which are partially enclosed by conducting surfaces free oscillations are possible, these oscillations dying away with a rapidity depending on the dimensions of the aperture by which the medium in the partially enclosed space communicates with that outside. When there are closed spaces in the medium occupied by different dielectric media, free oscillations are possible which decay with a rapidity depending on the difference between the velocities of radiation in the medium occupying a closed space and in the surrounding medium. These results can be utilised to form an idea of what is taking place when electrical disturbances are generated as in the case of Hertz' oscillator. The disturbances generated are propagated outwards from it with the velocity of radiation of the surrounding medium, and if this medium were everywhere the same and there were no partially enclosed spaces bounded by conducting surfaces, the disturbances would cease instantaneously to be propagated outwards when they ceased to be generated. This agrees with Bjerknes'[*] experiments where the disturbances are found to die away almost at once.

[*] *Annalen der Physik und Chemie*, Bd. 44, 1891.

CHAPTER VII.

PROPAGATION OF ELECTRICAL EFFECTS IN
MULTIPLY-CONNECTED SPACES.

42. IF there is a single perfect conductor whose surface is such that the space outside it is doubly connected, then resuming the argument of § 35 and considering the case where the wave fronts cut the surface of the conductor orthogonally, it follows that starting from any position of a wave front cutting the surface of the conductor in a closed curve A, and travelling with this wave front, the curve A will in no position become evanescent, and free permanent electrical oscillations will be possible, their periods being determined by the condition that, when the wave front has returned to the position from which it started, the wave is in the same phase. This result as well as that of § 35 can be deduced from the equations (6) of § 13. The result can be extended at once to the case where there are any number of conductors and the order of connexion of the space is any integer greater than two. The simplest case is that of a very thin wire in the form of a closed circuit; the free periods in this case are given by s/nV, where s is the length of the wire, V is the velocity of radiation in the medium outside the wire and n is any integer.

43. A solution can easily be obtained for the case of waves propagated in the direction of any number of parallel straight conductors. Let the axis of z be chosen in the direction of the conductors and let the transformation

$$\xi + \iota\eta = f(x + \iota y)$$

be determined so that the curves $\eta = \text{const.}$ include the boundaries of the cross sections of the conductors by any plane parallel to $z = 0$; the equations expressed in terms of the coordinates ξ, η, z are, § 36, the medium being supposed non-magnetic,

$$4\pi u = h_2 h_3 \left\{ \frac{\partial}{\partial \eta} \left(\frac{\gamma}{h_3} \right) - \frac{\partial}{\partial z} \left(\frac{\beta}{h_2} \right) \right\},$$

$$4\pi v = h_3 h_1 \left\{ \frac{\partial}{\partial z} \left(\frac{\alpha}{h_1} \right) - \frac{\partial}{\partial \xi} \left(\frac{\gamma}{h_3} \right) \right\},$$

$$4\pi w = h_1 h_2 \left\{ \frac{\partial}{\partial \xi} \left(\frac{\beta}{h_2} \right) - \frac{\partial}{\partial \eta} \left(\frac{\alpha}{h_1} \right) \right\},$$

$$-\frac{\partial \alpha}{\partial t} = h_2 h_3 \left\{ \frac{\partial}{\partial \eta} \left(\frac{Z}{h_3} \right) - \frac{\partial}{\partial z} \left(\frac{Y}{h_2} \right) \right\},$$

$$-\frac{\partial \beta}{\partial t} = h_3 h_1 \left\{ \frac{\partial}{\partial z} \left(\frac{X}{h_1} \right) - \frac{\partial}{\partial \xi} \left(\frac{Z}{h_3} \right) \right\},$$

$$-\frac{\partial \gamma}{\partial t} = h_1 h_2 \left\{ \frac{\partial}{\partial \xi} \left(\frac{Y}{h_2} \right) - \frac{\partial}{\partial \eta} \left(\frac{X}{h_1} \right) \right\}.$$

In this case the wave fronts are plane and are given by $z = \text{const.}$ Hence the equations become, remembering that

$$h_3 = 1, \quad h_1 = h_2 = J,$$

$$4\pi u = - J \frac{\partial}{\partial z} \left(\frac{\beta}{J} \right),$$

$$4\pi v = J \frac{\partial}{\partial z} \left(\frac{\alpha}{J} \right),$$

$$0 = \frac{\partial}{\partial \xi} \left(\frac{\beta}{J} \right) - \frac{\partial}{\partial \eta} \left(\frac{\alpha}{J} \right),$$

$$-\frac{\partial \alpha}{\partial t} = - J \frac{\partial}{\partial z} \left(\frac{Y}{J} \right),$$

$$-\frac{\partial \beta}{\partial t} = J \frac{\partial}{\partial z} \left(\frac{X}{J} \right),$$

$$0 = \frac{\partial}{\partial \xi} \left(\frac{Y}{J} \right) - \frac{\partial}{\partial \eta} \left(\frac{X}{J} \right).$$

From the last of these it follows that X and Y can be derived from a potential ψ, that is

$$X = J \frac{\partial \psi}{\partial \xi}, \qquad Y = J \frac{\partial \psi}{\partial \eta},$$

and the other equations become

$$\frac{J}{V^2} \frac{\partial^2 \psi}{\partial \xi \partial t} = - J \frac{\partial}{\partial z} \left(\frac{\beta}{J} \right),$$

$$\frac{J}{V^2} \frac{\partial^2 \psi}{\partial \eta \partial t} = J \frac{\partial}{\partial z} \frac{\alpha}{J} \right),$$

$$0 = \frac{\partial}{\partial \xi} \left(\frac{\beta}{J} \right) - \frac{\partial}{\partial \eta} \left(\frac{\alpha}{J} \right),$$

$$-\frac{\partial \alpha}{\partial t} = - J \frac{\partial^2 \psi}{\partial z \partial \eta},$$

$$-\frac{\partial \beta}{\partial t} = J \frac{\partial^2 \psi}{\partial z \partial \xi},$$

which are all satisfied if

$$\frac{1}{V^2} \frac{\partial^2 \psi}{\partial t^2} = \frac{\partial^2 \psi}{\partial z^2},$$

$$\frac{\partial^2 \psi}{\partial \xi^2} + \frac{\partial^2 \psi}{\partial \eta^2} = 0.$$

The boundary conditions are satisfied if ψ is constant over the boundary. The type of solution is then given by

$$\psi = A \eta e^{\iota \kappa (Vt \pm z)};$$

thus whenever the transformation can be found, the corresponding problem can be solved. For any number of thin wires the solution is

$$\psi = \Sigma A \log r \, . \, e^{\iota \kappa (Vt \pm z)},$$

where r is the distance from one of the wires and the summation is extended to all the wires. This result can be used to obtain the effect of a cylindrical conductor on the waves in the wire. For example, the effect of a circular cylindrical conductor, when there is one wire outside it parallel to it, is obtained by supposing the cylinder to be removed and another thin wire placed at the image of the former in the cylinder. The solution for the case where the medium is not the same throughout can also be obtained in two cases, viz. when the surfaces separating two different media are planes perpendicular to the wires, and when the surfaces separating them are cylindrical, their generators being parallel to the wires*.

44. It was shewn in § 13 that, if the magnetic force at each point of the surface of the perfect conductors were known, the electric and magnetic forces at each point of the medium could be expressed in the form of surface integrals taken over the surfaces of the conductors. When a conductor is a very thin wire, these surface integrals become, in the limit, line integrals taken along the wire and equations (6) § 13 take the form

$$X = \int \left(l_1 \frac{\partial^2}{\partial x^2} + m_1 \frac{\partial^2}{\partial x \partial y} + n_1 \frac{\partial^2}{\partial x \partial z} + \kappa^2 l_1 \right) L \frac{e^{-\iota\kappa r}}{r} \, ds,$$

$$Y = \int \left(l_1 \frac{\partial^2}{\partial y \partial x} + m_1 \frac{\partial^2}{\partial y^2} + n_1 \frac{\partial^2}{\partial y \partial z} + \kappa^2 m_1 \right) L \frac{e^{-\iota\kappa r}}{r} \, ds,$$

$$Z = \int \left(l_1 \frac{\partial^2}{\partial z \partial x} + m_1 \frac{\partial^2}{\partial z \partial y} + n_1 \frac{\partial^2}{\partial z^2} + \kappa^2 n_1 \right) L \frac{e^{-\iota\kappa r}}{r} \, ds,$$

where, to a factor, L is the line integral of the magnetic force taken round the section of the thin wire. Observing that l_1, m_1, n_1 are now the direction cosines of the tangent to the wire at a point (x_0, y_0, z_0) on it and that

$$\frac{\partial}{\partial x} \left(\frac{e^{-\iota\kappa r}}{r} \right) = - \frac{\partial}{\partial x_0} \left(\frac{e^{-\iota\kappa r}}{r} \right), \text{ etc.,}$$

the expressions for X, Y, Z become

$$X = \int \left[\frac{\partial}{\partial x} \left\{ - L \frac{\partial}{\partial s} \cdot \frac{e^{-\iota\kappa r}}{r} \right\} + \kappa^2 l_1 L \frac{e^{-\iota\kappa r}}{r} \right] ds,$$

$$Y = \int \left[\frac{\partial}{\partial y} \left\{ - L \frac{\partial}{\partial s} \cdot \frac{e^{-\iota\kappa r}}{r} \right\} + \kappa^2 m_1 L \frac{e^{-\iota\kappa r}}{r} \right] ds,$$

$$Z = \int \left[\frac{\partial}{\partial z} \left\{ - L \frac{\partial}{\partial s} \cdot \frac{e^{-\iota\kappa r}}{r} \right\} + \kappa^2 n_1 L \frac{e^{-\iota\kappa r}}{r} \right] ds,$$

* Cf. Lord Rayleigh, *Phil. Mag.* August, 1897.

which, since the wire forms a closed circuit, are equivalent to

$$X = \frac{\partial}{\partial x} \int \frac{e^{-\iota\kappa r}}{r} \frac{\partial L}{\partial s} \, ds + \kappa^2 \int l_1 L \, \frac{e^{-\iota\kappa r}}{r} \, ds,$$

$$Y = \frac{\partial}{\partial y} \int \frac{e^{-\iota\kappa r}}{r} \frac{\partial L}{\partial s} \, ds + \kappa^2 \int m_1 L \, \frac{e^{-\iota\kappa r}}{r} \, ds,$$

$$Z = \frac{\partial}{\partial z} \int \frac{e^{-\iota\kappa r}}{r} \frac{\partial L}{\partial s} \, ds + \kappa^2 \int n_1 L \, \frac{e^{-\iota\kappa r}}{r} \, ds.$$

It remains now to determine L to satisfy the boundary conditions at the surface of the wire. Assuming that the radius of the wire r_0 is small in comparison with the wave length, at a point on the wire

$$\int \frac{e^{-\iota\kappa r}}{r} \frac{\partial L}{\partial s} \, ds = - \, 2 \log r_0 \frac{\partial L}{\partial s} \, ,$$

$$\int \frac{e^{-\iota\kappa r}}{r} \, l_1 L \, ds = - \, 2 l_1 \log r_0 L, \, \ldots \ldots$$

where L, $\frac{\partial L}{\partial s}$, l_1, etc. are the values of these quantities at the point of the wire under consideration, the curvature of the wire at that point being continuous, and only the most important parts of the integrals being retained. In the case then of a single wire forming a closed circuit, the condition to be satisfied at any point is that

$$- \, 2 \log r_0 \left(\frac{\partial^2 L}{\partial s^2} + \kappa^2 L \right),$$

this being the component of the electric force along the wire, should vanish at every point of it. The component of the electric force tangential to the cross section of the wire obviously vanishes. The above condition requires

$$L = A e^{\pm \iota\kappa s},$$

and further, since after travelling round the circuit once L must return to the value from which it started, κs_0 must be a multiple of 2π, where s_0 is the length of the wire. The expressions for the components of the electric force at any point, when there is a single thin wire in the form of a closed circuit along which steady waves are travelling, are

$$X = \sum_{1}^{\infty} A_m \left\{ \frac{\partial}{\partial x} \int \iota \kappa_m e^{\iota \kappa_m (s-r)} \frac{ds}{r} + \kappa_m{}^2 \int l_1 e^{\iota \kappa_m (s-r)} \frac{ds}{r} \right\},$$

$$Y = \sum_{1}^{\infty} A_m \left\{ \frac{\partial}{\partial y} \int \iota \kappa_m e^{\iota \kappa_m (s-r)} \frac{ds}{r} + \kappa_m{}^2 \int m_1 e^{\iota \kappa_m (s-r)} \frac{ds}{r} \right\},$$

$$Z = \sum_{1}^{\infty} A_m \left\{ \frac{\partial}{\partial z} \int \iota \kappa_m e^{\iota \kappa_m (s-r)} \frac{ds}{r} + \kappa_m{}^2 \int n_1 e^{\iota \kappa_m (s-r)} \frac{ds}{r} \right\},$$

where $A_m = C_m e^{\iota \kappa_m Vt}$, C_m is a constant, and $\kappa_m = 2m\pi/s_0$, m being an integer. These expressions hold accurately in the limit when the cross section of the wire is indefinitely diminished, provided that the curvature is at all points of the circuit continuous.

45. When the wire is not supposed to be indefinitely thin, these expressions will still be a first approximation to the accurate ones. The component of the electric force normal to the surface of the wire at any point along a cross section has a term which involves $\log r_0$ as a factor, where r_0 is the radius of the thin wire. The next term in order of magnitude cannot be obtained from the line integral; to obtain it the surface integrals of equations (6) § 13 must be used, as to this order the magnetic force cannot be taken constant along a cross section. The result will be a term which, when r_0 is small compared with the radius of curvature of the wire, is negligible in comparison with the first term. Thus the result given above § 44 is applicable in the case of a thin wire whose radius r_0 is small compared with the radius of curvature of the wire and with the wave length of the oscillations*.

46. When electric waves are incident on a closed perfectly conducting wire the surface integrals must also be used to express the electric forces, as in this case the line integral of

---

* The case of a wire in the form of a circle has been discussed by Pocklington, *Camb. Phil. Soc. Proc.* 1892; he finds a complex value for the fundamental period when the cross section is finite; this has arisen from his expressing the electric forces by means of line integrals, which is illegitimate when the cross section is finite; in the limit when the cross section is negligible his results agree with those which would follow from the methods set forth above.

the magnetic force taken round a curve looping the wire is evanescent with the curve. Thus L is zero so far as it depends on the incident waves, and if the cross section of the wire be supposed to be circular the magnetic force at any point on it will be expressible as a Fourier series involving sines of multiples of the azimuth. The induced vibrations will be most definite when the coefficients in this Fourier series are periodic of the same period as a system of steady waves belonging to the circuit, and then waves will travel along the wire with the velocity of radiation of the medium outside the wire just as in the case of the waves belonging to the circuit. These induced vibrations have been utilised by Blondlot* to determine experimentally the circumstances of the propagation of electric waves along wires.

In the above the wires have been assumed to be perfectly conducting; the effect of imperfect conduction in a straight wire has been discussed by Sommerfeld†, the result arrived at being that the effect is very small. The investigation assumes that the energy of the waves is dissipated according to Ohm's law, which although true for very long waves is unlikely to hold for short waves; evidence for any theory as to how the propagation of waves of short wave length is affected by resistance has yet to be obtained. The results of § 44 for a single circuit can now, in virtue of what has been stated as to induced vibrations in a circuit, be extended to any number of circuits, the result being a sum of expressions of the kind there given. This can be used to construct solvable cases as to the effect of conductors on waves in a closed circuit just as in the case of potential theory.

* *Journal de Physique*, Vol. x.
† *Annalen der Physik und Chemie*, Bd. 67, 1899.

CHAPTER VIII.

RADIATION.

47. It has already been shewn in Chapter IV. that Maxwell's second expression for the electrokinetic energy, which expresses it as one half of the vector product of the magnetic induction and the magnetic force per unit volume, is, in any case where electrical effects are being propagated, inconsistent with his first expression, which is the one built up from the results of observation. The analytical expression for the change of electric energy inside any closed surface, or for the rate of transfer of energy across the surface, will therefore be different from that given by Poynting*, who has in his investigation taken the second expression for the measure of the electrokinetic energy.

The amount of electric energy inside a closed surface, which does not enclose any conductors or electric charges, is by § 23

$$\tfrac{1}{2} \iiint \{Ff + Gg + Hh + Xf + Yg + Zh\}\, dx\,dy\,dz,$$

where F, G, H are the components of the electrokinetic momentum, X, Y, Z are the components of the electric force, f, g, h are the components of the electric displacement at the point x, y, z, and the integral is taken throughout the space enclosed by the surface. The time rate of change of this quantity is

$$\tfrac{1}{2} \iiint \{F\ddot{f} + G\ddot{g} + H\ddot{h} + \dot{F}\dot{f} + \dot{G}\dot{g} + \dot{H}\dot{h}$$
$$+ \dot{X}f + \dot{Y}g + \dot{Z}h + X\dot{f} + Y\dot{g} + Z\dot{h}\}\, dx\,dy\,dz;$$

* *Phil. Trans.*, 1884.

now $\qquad X = 4\pi V^2 f, \quad Y = 4\pi V^2 g, \quad Z = 4\pi V^2 h,$

and $\qquad X = -\dfrac{\partial F}{\partial t} - \dfrac{\partial \phi}{\partial x}, \quad Y = -\dfrac{\partial G}{\partial t} - \dfrac{\partial \phi}{\partial y}, \quad Z = -\dfrac{\partial H}{\partial t} - \dfrac{\partial \phi}{\partial z},$

whence the time rate of increase of the energy inside the surface is

$$\tfrac{1}{2} \iiint \left\{ F\ddot{f} + G\ddot{g} + H\ddot{h} - \ddot{F}f - \ddot{G}g - \ddot{H}h \right.$$
$$\left. - \dfrac{\partial \phi}{\partial x} \dot{f} - \dfrac{\partial \phi}{\partial y} \dot{g} - \dfrac{\partial \phi}{\partial z} \dot{h} \right\} dx\,dy\,dz.$$

In the case of the propagation of waves of one definite period, if there are no electric charges present except those which are oscillating, ϕ will be zero and the above expression will vanish identically*, so that, in this case, the amount of energy inside a closed surface, which does not enclose any conductors or electric charges, is constant. If in the same case there are present electric charges in addition to the oscillating ones, the time rate of increase of the energy inside is

$$- \iiint \left\{ \dfrac{\partial \phi}{\partial x} \dot{f} + \dfrac{\partial \phi}{\partial y} \dot{g} + \dfrac{\partial \phi}{\partial z} \dot{h} \right\} dx\,dy\,dz,$$

whence the total change for a complete period is zero. When there are present waves of different periods, the total change for a time, which is a multiple of all the periods, vanishes, and in every case the energy travels with the waves.

48. When there are electric charges inside the surface, to obtain the time rate of change of the energy inside the surface, the rate of change due to the motions of the moving charges must be added. Remembering that the components of the mechanical force acting on the unit of volume are

$$(X + \gamma q - \beta r)\,\rho, \quad (Y + \alpha r - \gamma p)\,\rho, \quad (Z + \beta p - \alpha q)\,\rho,$$

where ρ is the density per unit volume of electric charge, p, q, r are the components of the velocity of the charge, and α, β, γ are the components of the magnetic force at the point,

* See Appendix C.

the time rate of increase of the amount of electric energy inside
the surface is

$$\frac{d}{dt} \cdot \frac{1}{2} \iiint \{Fu + Gv + Hw + Xf + Yg + Zh\} \, dx \, dy \, dz$$
$$+ \iiint \{(X + \gamma q - \beta r) \, \rho p + (Y + \alpha r - \gamma p) \, \rho q$$
$$+ (Z + \beta p - \alpha q) \, \rho r\} \, dx \, dy \, dz,$$

where u, v, w are the components of the total current. This
expression is equivalent to

$$\frac{d}{dt} \cdot \frac{1}{2} \iiint \{Fu + Gv + Hw\} \, dx \, dy \, dz + \iiint \{X \, (\dot{f} + \rho p) + Y \, (\dot{g} + \rho q)$$
$$+ Z \, (\dot{h} + \rho r)\} \, dx \, dy \, dz,$$

that is, to

$$\frac{d}{dt} \cdot \frac{1}{2} \iiint \{Fu + Gv + Hw\} \, dx \, dy \, dz + \iiint \{Xu + Yv + Zw\} \, dx \, dy \, dz.$$

Writing for the current strengths their equivalents in terms
of the magnetic force, the above expression becomes

$$\frac{1}{8\pi} \frac{d}{dt} \iiint \left[F \left(\frac{\partial \gamma}{\partial y} - \frac{\partial \beta}{\partial z} \right) + G \left(\frac{\partial \alpha}{\partial z} - \frac{\partial \gamma}{\partial x} \right) + H \left(\frac{\partial \beta}{\partial x} - \frac{\partial \alpha}{\partial y} \right) \right] dx \, dy \, dz$$
$$+ \frac{1}{4\pi} \iiint \left[X \left(\frac{\partial \gamma}{\partial y} - \frac{\partial \beta}{\partial z} \right) + Y \left(\frac{\partial \alpha}{\partial z} - \frac{\partial \gamma}{\partial x} \right) + Z \left(\frac{\partial \beta}{\partial x} - \frac{\partial \alpha}{\partial y} \right) \right] dx \, dy \, dz,$$

that is, integrating by parts

$$\frac{1}{8\pi} \frac{d}{dt} \iint [F \, (m\gamma - n\beta) + G \, (n\alpha - l\gamma) + H \, (l\beta - m\alpha)] \, dS'$$

$$+ \frac{1}{4\pi} \iint [X \, (m\gamma - n\beta) + Y \, (n\alpha - l\gamma) + Z \, (l\beta - m\alpha)] \, dS'$$

$$+ \frac{1}{8\pi} \frac{d}{dt} \iiint \left[\alpha \left(\frac{\partial H}{\partial y} - \frac{\partial G}{\partial z} \right) + \beta \left(\frac{\partial F}{\partial z} - \frac{\partial H}{\partial x} \right) + \gamma \left(\frac{\partial G}{\partial x} - \frac{\partial F}{\partial y} \right) \right] dx \, dy \, dz$$

$$+ \frac{1}{4\pi} \iiint \left[\alpha \left(\frac{\partial Z}{\partial y} - \frac{\partial Y}{\partial z} \right) + \beta \left(\frac{\partial X}{\partial z} - \frac{\partial Z}{\partial x} \right) + \gamma \left(\frac{\partial Y}{\partial x} - \frac{\partial X}{\partial y} \right) \right] dx \, dy \, dz.$$

Now the volume integral in the above expression is equivalent to

$$\frac{1}{8\pi} \frac{d}{dt} \iiint (\alpha^2 + \beta^2 + \gamma^2) \, dx \, dy \, dz$$
$$- \frac{1}{4\pi} \iiint \left(\alpha \frac{d\alpha}{dt} + \beta \frac{d\beta}{dt} + \gamma \frac{d\gamma}{dt} \right) dx \, dy \, dz,$$

which vanishes. Hence the rate, at which energy is transferred across the surface into the space inside it, is

$$\frac{1}{8\pi}\frac{d}{dt}\iint [F(m\gamma - n\beta) + G(n\alpha - l\gamma) + H(l\beta - m\alpha)]\, dS'$$

$$+ \frac{1}{4\pi}\iint [X(m\gamma - n\beta) + Y(n\alpha - l\gamma) + Z(l\beta - m\alpha)]\, dS',$$

where l, m, n are the direction cosines of the normal to the surface at the point drawn outwards. If ϵ is the angle, which the normal to the surface makes with the direction which is perpendicular to the directions of the magnetic and electric forces, the angle between these two directions being θ, and ϵ_1 is the angle, which the normal to the surface makes with the direction which is perpendicular to the directions of the magnetic force and the electrokinetic momentum, the angle between these two directions being θ_1, the amount of energy, which in time dt crosses the element of area dS' of the surface at the point, is

$$\left[\frac{1}{8\pi}\frac{d}{dt}MH\sin\theta_1\cos\epsilon_1 + \frac{1}{4\pi}EH\sin\theta\cos\epsilon\right] dt\, dS',$$

where M is the resultant electrokinetic momentum, H is the resultant magnetic force, and E is the resultant electric force.

The first term in the above, when summed for a complete period, when there are oscillations of only one type, or for a time, which is an integral multiple of all the periods belonging to the waves which in any case are present, vanishes, and there will remain the time integral of the second term, which is Poynting's expression for the rate of transfer of the energy; so that, for a time after which the state of affairs inside the surface is the same as at the beginning, the expression for the amount of energy, which crosses the surface, will be the same, whether Poynting's expression for the rate of transfer or the above be taken; in any other case they will be different.

49. A case of great practical importance is that of a simple Hertzian oscillator, which will now be discussed. Taking the axis of z in the direction of the axis of the oscillator, the components of the electric force can be written in the form

$$X = -A \frac{\partial^2 S}{\partial x \partial z}, \quad Y = -A \frac{\partial^2 S}{\partial y \partial z}, \quad Z = A \left(\frac{\partial^2}{\partial x^2} + \frac{\partial^2}{\partial y^2} \right) S,$$

where

$$S = \frac{\sin \kappa (r - Vt)}{r},$$

r is the distance of the point x, y, z from the centre of the oscillator, $2\pi/\kappa$ is the wave length of the oscillations, V is the velocity of radiation in the surrounding medium, and A is a constant which can be determined as follows. In the immediate neighbourhood of the oscillator r is a very small quantity, and can be taken so small that it is negligible in comparison with the wave length; the components of the electric force very close to the oscillator are therefore

$$X = A \frac{\partial^2}{\partial x \partial z} \frac{1}{r} \sin \kappa Vt, \quad Y = A \frac{\partial^2}{\partial y \partial z} \frac{1}{r} \sin \kappa Vt,$$

$$Z = -A \left(\frac{\partial^2}{\partial x^2} + \frac{\partial^2}{\partial y^2} \right) \frac{1}{r} \sin \kappa Vt,$$

the last of which is equivalent to

$$Z = A \frac{\partial^2}{\partial z^2} \cdot \frac{1}{r} \sin \kappa Vt,$$

whence, close up to the oscillator, the components of the electric force are derivable from a potential

$$-A \frac{\partial}{\partial z} \cdot \frac{1}{r} \cdot \sin \kappa Vt,$$

and therefore

$$A = ELV^2,$$

where E is an electric charge and L is a length, everything being supposed expressed in electromagnetic units. The components of the electrokinetic momentum are given by

$$F = \frac{A}{\kappa V} \frac{\partial^2 C}{\partial x \partial z}, \quad G = \frac{A}{\kappa V} \frac{\partial^2 C}{\partial y \partial z}, \quad H = -\frac{A}{\kappa V} \left(\frac{\partial^2}{\partial x^2} + \frac{\partial^2}{\partial y^2} \right) C,$$

where
$$C = \frac{\cos \kappa \, (r - Vt)}{r} \, ;$$
and hence, using the relations

$$\alpha = \frac{\partial H}{\partial y} - \frac{\partial G}{\partial z}, \quad \beta = \frac{\partial F}{\partial z} - \frac{\partial H}{\partial x}, \quad \gamma = \frac{\partial G}{\partial x} - \frac{\partial F}{\partial y},$$

it follows that

$$\alpha = - \frac{A}{\kappa V} \frac{\partial}{\partial y} \left\{ \frac{\partial^2}{\partial x^2} + \frac{\partial^2}{\partial y^2} + \frac{\partial^2}{\partial z^2} \right\} C,$$

$$\beta = \frac{A}{\kappa V} \frac{\partial}{\partial x} \left\{ \frac{\partial^2}{\partial x^2} + \frac{\partial^2}{\partial y^2} + \frac{\partial^2}{\partial z^2} \right\} C,$$

$$\gamma = 0,$$

that is,

$$\alpha = - \frac{A}{\kappa V^3} \frac{\partial^3 C}{\partial t^2 \partial y}, \quad \beta = \frac{A}{\kappa V^3} \frac{\partial^3 C}{\partial t^2 \partial x}, \quad \gamma = 0,$$

or

$$\alpha = \frac{A \kappa}{V} \frac{\partial C}{\partial y}, \quad \beta = - \frac{A \kappa}{V} \frac{\partial C}{\partial x}, \quad \gamma = 0.$$

The rate at which energy flows into any closed space across its bounding surface S' is

$$\frac{1}{8\pi} \frac{d}{dt} \iint [F (m\gamma - n\beta) + G (n\alpha - l\gamma) + H (l\beta - m\alpha)] \, dS'$$

$$+ \frac{1}{4\pi} \iint [X (m\gamma - n\beta) + Y (n\alpha - l\gamma) + Z (l\beta - m\alpha)] \, dS',$$

which in this case is

$$\frac{1}{8\pi} \iint [F (m\dot{\gamma} - n\dot{\beta}) - \dot{F} (m\gamma - n\beta) + G (n\dot{\alpha} - l\dot{\gamma}) - \dot{G} (n\alpha - l\gamma)$$
$$+ H (l\dot{\beta} - m\dot{\alpha}) - \dot{H} (l\beta - m\alpha)] \, dS'.$$

Now, there being only one period of oscillation, and no free non-oscillating charge present, it follows by § 47, that the above expression is the same for all surfaces, which enclose the oscillator. Taking as the surface any sphere of radius r with its centre at the oscillator, the direction of the normal at any point is given by

$$l = \frac{x}{r}, \quad m = \frac{y}{r}, \quad n = \frac{z}{r},$$

and therefore

$$F(m\dot\gamma - n\dot\beta) - \dot F(m\gamma - n\beta) = \frac{A^2\kappa}{V}\left[\frac{z}{r}\frac{\partial^2 C}{\partial x \partial z}\frac{\partial S}{\partial x} - \frac{z}{r}\frac{\partial^2 S}{\partial x \partial z}\frac{\partial C}{\partial x}\right],$$

$$G(n\dot\alpha - l\dot\gamma) - \dot G(n\alpha - l\gamma) = \frac{A^2\kappa}{V}\left[\frac{z}{r}\frac{\partial^2 C}{\partial y \partial z}\frac{\partial S}{\partial y} - \frac{z}{r}\frac{\partial^2 S}{\partial y \partial z}\frac{\partial C}{\partial y}\right],$$

$$H(l\dot\beta - m\dot\alpha) - \dot H(l\beta - m\alpha) = \frac{A^2\kappa}{V}\left[\left(\frac{\partial^2}{\partial x^2} + \frac{\partial^2}{\partial y^2}\right)C\left(\frac{x}{r}\frac{\partial S}{\partial x} + \frac{y}{r}\frac{\partial S}{\partial y}\right)\right.$$
$$\left. - \left(\frac{\partial^2}{\partial x^2} + \frac{\partial^2}{\partial y^2}\right)S\left(\frac{x}{r}\frac{\partial C}{\partial x} + \frac{y}{r}\frac{\partial C}{\partial y}\right)\right].$$

Hence the rate at which energy flows into the space inside the sphere is

$$\frac{A^2\kappa}{8\pi V}\iint\left[\left\{\frac{zx}{r^2}\frac{\partial^2 C}{\partial x \partial z} + \frac{zy}{r^2}\frac{\partial^2 C}{\partial y \partial z} + \frac{x^2+y^2}{r^2}\left(\frac{\partial^2}{\partial x^2} + \frac{\partial^2}{\partial y^2}\right)C\right\}\frac{\partial S}{\partial r}\right.$$
$$\left. - \left\{\frac{zx}{r^2}\frac{\partial^2 S}{\partial x \partial z} + \frac{zy}{r^2}\frac{\partial^2 S}{\partial y \partial z} + \frac{x^2+y^2}{r^2}\left(\frac{\partial^2}{\partial x^2} + \frac{\partial^2}{\partial y^2}\right)S\right\}\frac{\partial C}{\partial r}\right]dS,$$

which is equivalent to

$$\frac{A^2\kappa}{8\pi V}\iint\left[\left\{\left(\frac{z^2x^2}{r^4} + \frac{z^2y^2}{r^4} + \frac{(x^2+y^2)^2}{r^4}\right)\frac{\partial^2 C}{\partial r^2} - \left(\frac{z^2x^2}{r^5} + \frac{z^2y^2}{r^5}\right.\right.\right.$$
$$\left.\left. + \frac{x^2+y^2}{r^2}\left(\frac{x^2+y^2}{r^3} - \frac{2}{r}\right)\right)\frac{\partial C}{\partial r}\right\}\frac{\partial S}{\partial r} - \left\{\left(\frac{z^2x^2}{r^4} + \frac{z^2y^2}{r^4} + \frac{(x^2+y^2)^2}{r^4}\right)\frac{\partial^2 S}{\partial r^2}\right.$$
$$\left.\left. - \left(\frac{z^2x^2}{r^5} + \frac{z^2y^2}{r^5} + \frac{x^2+y^2}{r^2}\left(\frac{x^2+y^2}{r^3} - \frac{2}{r}\right)\right)\frac{\partial S}{\partial r}\right\}\frac{\partial C}{\partial r}\right]dS,$$

that is, to

$$\frac{A^2\kappa}{8\pi V}\iint\left[\frac{x^2+y^2}{r^2}\frac{\partial^2 C}{\partial r^2}\frac{\partial S}{\partial r} - \frac{x^2+y^2}{r^2}\frac{\partial^2 S}{\partial r^2}\frac{\partial C}{\partial r}\right]dS,$$

or to

$$\frac{A^2\kappa}{8\pi V}\iint\left[\left(-\kappa^2 C + \frac{2\kappa}{r}S + \frac{2}{r^2}C\right)\left(\kappa C - \frac{1}{r}S\right)\right.$$
$$\left. - \left(-\kappa^2 S - \frac{2\kappa}{r}C + \frac{2}{r^2}S\right)\left(\kappa S + \frac{1}{r}C\right)\right]\frac{x^2+y^2}{r^2}dS,$$

which is

$$-\frac{A^2\kappa^4}{8\pi V}\iint\frac{x^2+y^2}{r^4}dS$$

or

$$-\frac{A^2\kappa^4}{3V}.$$

Therefore the energy is radiated out from the oscillator at a steady rate of amount $\dfrac{A^2\kappa^4}{3V}$, that is of amount $\dfrac{16\pi^4E^2L^2V^3}{3\lambda^4}$ in electromagnetic units, where λ is the wave length of the oscillations, and E is the maximum charge at an end of the oscillator. The rate of radiation in electrostatic units is

$$\frac{16\pi^4E^2L^2V}{3\lambda^4},$$

where EL is the maximum electric moment of the oscillator in electrostatic units. The amount radiated out in a half period is $\dfrac{8\pi^4E^2L^2}{3\lambda^3}$ in electrostatic units, which is the result given by Hertz*.

50. The rate of decay of the oscillations, sent out by an oscillator, can be obtained very approximately from the above result. Taking the case of Hertz' oscillator, which consists of two equal conducting spheres of radius R, the distance between their centres being l, the amount of electric energy in the oscillator is E^2/R†, approximately, where E is the maximum charge on either sphere, and when the supply of energy is stopped this amount remains to be radiated outwards. The rate of radiation of energy is $\dfrac{16\pi^4E^2l^2V}{3\lambda^4}$ per unit time, where λ is the wave length of the oscillations, and therefore

$$\frac{dW}{dt} = -\frac{16\pi^4E^2l^2V}{3\lambda^4},$$

approximately, when the supply of energy has been stopped, the rate having been calculated on the assumption that the amount of energy necessary to maintain a steady radiation is being supplied. Therefore

$$\frac{dW}{dt} = -\frac{16\pi^4l^2RV}{3\lambda^4}W,$$

that is $W = e^{-kt}W_0,$

where $k = 16\pi^4l^2RV/3\lambda^4,$

---

* *Electric Waves*, English Trans. p. 150.
† Hertz, *Electric Waves*, loc. cit.

and the time which elapses, before the amplitude of the oscillations falls to $1/e$ of its initial value, is $2/k$, that is $3\lambda^4/8\pi^4 l^2 R V$.

For similar oscillators, it follows that the time which elapses is the same number of oscillations, so that, for oscillators giving oscillations of short wave length, the decay is more rapid than for similar oscillators giving oscillations of greater wave length, a result observed by Hertz. The oscillator, used by Hertz in most of his experiments, was approximately in tune with a resonator whose fundamental wave length was 560 cm.; the wave length of the oscillations, sent out by the oscillator, was somewhat shorter, as in order to make the resonator have the same wave length small pieces of metal had to be soldered to the ends of the resonator, the effect of which, as will appear later, is to diminish the wave length of the resonator. The radius of either sphere of the oscillator was 15 cm., and the distance apart of their centres 100 cm., so that the time which elapses before the amplitude of the oscillations falls to $1/e$ of its initial value is somewhat less than $\dfrac{6 \times 560^3}{16\pi^4 \times 100^2 \times 15} \dfrac{\lambda}{V}$, which is $4\cdot4\,T$, where T is the time of a complete oscillation. The result obtained by Bjerknes* experimentally for a somewhat similar oscillator is $3\cdot8\,T$.

51. For some purposes it is important to know the density of electric energy at any point due to an oscillator. The density per unit volume of the energy is

$$\tfrac{1}{2}\left(F\dot{f} + G\dot{g} + H\dot{h} + Xf + Yg + Zh\right),$$

which, on substituting the values of F, G, H etc. given above § 49, becomes

$$\frac{A^2}{8\pi V^2}\left[\frac{(x^2+y^2)\,z^2}{r^2}\left(\frac{\partial}{\partial r}\frac{1}{r}\frac{\partial C}{\partial r}\right)^2 + \left(\frac{2}{r}\frac{\partial C}{\partial r} + \frac{x^2+y^2}{r}\frac{\partial}{\partial r}\cdot\frac{1}{r}\frac{\partial C}{\partial r}\right)^2\right.$$
$$\left. + \frac{(x^2+y^2)\,z^2}{r^2}\left(\frac{\partial}{\partial r}\frac{1}{r}\frac{\partial S}{\partial r}\right)^2 + \left(\frac{2}{r}\frac{\partial S}{\partial r} + \frac{x^2+y^2}{r}\frac{\partial}{\partial r}\cdot\frac{1}{r}\frac{\partial S}{\partial r}\right)^2\right],$$

* *Annalen der Physik und Chemie*, Bd. 44, 1891.

that is

$$\frac{A^2}{8\pi V^2}\left[(x^2+y^2)\left\{\left(\frac{1}{r}\frac{\partial^2 C}{\partial r^2}-\frac{1}{r^2}\frac{\partial C}{\partial r}\right)^2+\left(\frac{1}{r}\frac{\partial^2 S}{\partial r^2}-\frac{1}{r^2}\frac{\partial S}{\partial r}\right)^2\right.\right.$$
$$+\frac{4}{r^2}\frac{\partial C}{\partial r}\left(\frac{1}{r}\frac{\partial^2 C}{\partial r^2}-\frac{1}{r^2}\frac{\partial C}{\partial r}\right)+\frac{4}{r^2}\frac{\partial S}{\partial r}\left(\frac{1}{r}\frac{\partial^2 S}{\partial r^2}-\frac{1}{r^2}\frac{\partial S}{\partial r}\right)\right\}$$
$$\left.+\frac{4}{r^2}\left(\frac{\partial C}{\partial r}\right)^2+\frac{4}{r^2}\left(\frac{\partial S}{\partial r}\right)^2\right],$$

or

$$\frac{A^2}{8\pi V^2}\left[\frac{x^2+y^2}{r^2}\left\{\left(\frac{\partial^2 C}{\partial r^2}+\frac{1}{r}\frac{\partial C}{\partial r}\right)^2+\left(\frac{\partial^2 S}{\partial r^2}+\frac{1}{r}\frac{\partial S}{\partial r}\right)^2-\frac{4}{r^2}\left(\frac{\partial C}{\partial r}\right)^2-\frac{4}{r^2}\left(\frac{\partial S}{\partial r}\right)^2\right\}\right.$$
$$\left.+\frac{4}{r^2}\left(\frac{\partial C}{\partial r}\right)^2+\frac{4}{r^2}\left(\frac{\partial S}{\partial r}\right)^2\right],$$

which is

$$\frac{A^2}{8\pi V^2}\left[\frac{x^2+y^2}{r^2}\left\{\left(\frac{1}{r}\frac{\partial C}{\partial r}+\kappa^2 C\right)^2+\left(\frac{1}{r}\frac{\partial S}{\partial r}+\kappa^2 S\right)^2\right.\right.$$
$$\left.\left.-\frac{4}{r^2}\left(\frac{\partial C}{\partial r}\right)^2-\frac{4}{r^2}\left(\frac{\partial S}{\partial r}\right)^2\right\}+\left(\frac{4}{r^2}\frac{\partial C}{\partial r}\right)^2+\left(\frac{4}{r^2}\frac{\partial S}{\partial r}\right)^2\right];$$

now

$$C^2+S^2=\frac{1}{r^2},$$

$$C\frac{\partial C}{\partial r}+S\frac{\partial S}{\partial r}=-\frac{1}{r^3},$$

$$\left(\frac{\partial C}{\partial r}\right)^2+\left(\frac{\partial S}{\partial r}\right)^2=\frac{\kappa^2}{r^2}+\frac{1}{r^4},$$

hence the density per unit volume at the point x, y, z is

$$\frac{A^2}{8\pi V^2}\left[\frac{x^2+y^2}{r^2}\left\{\frac{\kappa^4}{r^2}-\frac{5\kappa^2}{r^4}-\frac{3}{r^6}\right\}+\frac{4\kappa^2}{r^4}+\frac{4}{r^6}\right],$$

which is equivalent to

$$\frac{E^2 l^2}{8\pi}\left[\left(\frac{\kappa^4}{r^2}-\frac{5\kappa^2}{r^4}-\frac{3}{r^6}\right)\sin^2\theta+\frac{4\kappa^2}{r^4}+\frac{4}{r^6}\right].$$

At a distance r from the origin, great compared with the wave length, the density is $\dfrac{E^2 l^2 \kappa^4 \sin^2\theta}{8\pi r^2}$, that is $\dfrac{2\pi^3 E^2 l^2 \sin^2\theta}{r^2\lambda^4}$, and for any number of the same oscillators, orientated indifferently, placed at the origin, the density at a distance r will be $B\dfrac{E^2 l^2}{r^2\lambda^4}$, where B is some number.

52. If there are in a space a number of ions describing orbits*, the wave lengths belonging to them being all the same, and the space be supposed to be contracted in such a way, that all lengths are diminished in the same ratio, the orbits in the contracted space will have their linear dimensions diminished in this ratio, and the wave lengths will also be diminished in the same ratio. Hence, denoting the density of the energy at any point in the original space by C/λ^4, the density in the new space at the corresponding point will be C/λ'^4, where λ' is the wave length appropriate to the new space. Now unit volume in the original space becomes a volume λ'^3/λ^3 in the new space, therefore the amount of work, per unit volume of the original space, which has to be done, to pass from the original state to the new state, is $C/\lambda'\lambda^3 - C/\lambda^4$, and therefore, if T and T' are the temperatures corresponding to the original and new states respectively,

$$\frac{T' - T}{T} = \left(\frac{C}{\lambda'\lambda^3} - \frac{C}{\lambda^4}\right)\Big/\frac{C}{\lambda^4},$$

whence λT is constant, and the density of the energy is proportional to T^4, which results have been previously obtained†. Further, if energy is being radiated away from the distribution of ions, the rate of this radiation is (§ 49) proportional to λ^{-4} and therefore to T^4.

53. The condition that a group of ions should be permanent, that is that no energy is radiated away from the group, can be obtained as follows. The components of the magnetic force due to the revolving ions are given by‡

$$\alpha = \frac{\partial^2}{\partial y \partial t}\Sigma\frac{e\zeta}{r} - \frac{\partial^2}{\partial z \partial t}\Sigma\frac{e\eta}{r},$$

$$\beta = \frac{\partial^2}{\partial z \partial t}\Sigma\frac{e\xi}{r} - \frac{\partial^2}{\partial x \partial t}\Sigma\frac{e\zeta}{r},$$

$$\gamma = \frac{\partial^2}{\partial x \partial t}\Sigma\frac{e\eta}{r} - \frac{\partial^2}{\partial y \partial t}\Sigma\frac{e\xi}{r},$$

---

* An ion describing an orbit can be replaced by a number of oscillators (Appendix C).

† Wien, *Berlin. Sitzungsberichte*, 1893; Larmor, *Aether and Matter*, p. 137, 1900.

‡ Appendix C.

where ξ, η, ζ are the displacements of an ion at the time $t - \dfrac{r}{V}$, these displacements being supposed small, and the summation being extended to all the ions of the group. Now the condition, that there should be no radiation of energy away from the group, is that the integral

$$\frac{1}{8\pi} \frac{d}{dt} \iint \{F (m\gamma - n\beta) + G (n\alpha - l\gamma) + H (l\beta - m\alpha)\} \, dS$$

$$+ \frac{1}{4\pi} \iint \{X (m\gamma - n\beta) + Y (n\alpha - l\gamma) + Z (l\beta - m\alpha)\} \, dS,$$

taken over the surface of a sphere at a great distance from the group, should vanish; the condition will be satisfied if the terms in α, β, γ, $\dot{\alpha}$, $\dot{\beta}$, $\dot{\gamma}$, involving the inverse power of the distance, vanish, and this requires

$$\Sigma e\bar{\xi} = 0, \quad \Sigma e\bar{\eta} = 0, \quad \Sigma e\bar{\zeta} = 0,$$

where $\bar{\xi}, \bar{\eta}, \bar{\zeta}$ are the maximum displacements of an ion in the directions of the axes of reference.

54. The force, which one permanent group exerts on another, can be obtained from the preceding results. Taking first the case where there is no free electricity, the condition $\Sigma e = 0$ will be satisfied for each group in addition to the conditions given above. The conditions $\Sigma e\bar{\xi} = \Sigma e\bar{\eta} = \Sigma e\bar{\zeta} = 0$ being satisfied, the lowest power of $1/r$, which can occur in the components of the magnetic force due to a permanent group at a distance r from it, is $1/r^3$, and, the condition $\Sigma e = 0$ being also satisfied for the group, the lowest power of $1/r$ which can occur in the components of the electric force due to it is $1/r^3$. The components of the force, exerted by this group on any other group, are

$$\Sigma Xe + \Sigma e\dot{\eta}\gamma - \Sigma e\dot{\zeta}\beta, \quad \Sigma Ye + \Sigma e\dot{\zeta}\alpha - \Sigma e\dot{\xi}\gamma, \quad \Sigma Ze + \Sigma e\dot{\xi}\beta - \Sigma e\dot{\eta}\alpha,$$

where X, Y, Z, α, β, γ are the components of the electric and magnetic forces due to the first group, and the summation extends to all the ions of the second group. If the second group is permanent and there is no free electricity belonging to it, the lowest power of $1/r$, which can occur in the above expressions for the components of the force, is $1/r^4$ and the order of magnitude of this part of the force is $e^2 l^2/r^4$, where l is the diameter of an orbit, so that only the groups, which are in the near neighbourhood of any given group, exert a sensible force on it. When there is free electricity belonging to one of the two groups, the lowest power of $1/r$, which can occur in the expressions for the components of the force, is $1/r^3$, and the order of magnitude of this part is $e^2 l/r^3$. When there is free electricity belonging to both groups, the lowest power of $1/r$, which can occur in the expressions for the force between them, is $1/r^2$, the order of magnitude of this part is e^2/r^2 and its direction is that of the line joining them.

It follows from the preceding that, in any material medium in which there is no free electricity present, the forces, which the groups of ions, which constitute the material medium, exert on each other are only sensible at very small distances. When there is free electricity present in the material medium, the part of the forces between the groups of ions, which is sensible at a finite distance, consists of forces between pairs of groups which possess free electricity, the force between any pair varying inversely as the square of their distance apart, its direction being that of the line joining them, and being repulsive or attractive according as the free electricity in the two groups consists of an excess of ions of the same or opposite signs. In different material media the configurations of the groups will be different, and, between any two different material media in contact, there will be a transition layer in which there are groups belonging to both kinds of matter. The absolute values of the forces between groups of different kinds will be different from that between groups of the same kind, and to this is to be ascribed the phenomena of capillarity and other phenomena associated with the contact of different media. So long as the groups are permanent, whether there is free electricity present or not, there is no radiation of energy away from the system; this only takes place when the groups are being broken up to form new groups, and the intensity of this radiation for a definite wave length varies inversely as the fourth power of the wave length, as has been established above.

55. The results of the previous chapter, as to the permanence of waves associated with closed circuits, can also be established from considerations relative to the radiation of energy from the circuits. Taking the case of waves travelling along an infinitely extended straight wire, the expressions for the components of the electric force at any point are (§ 43)

$$X = \frac{Ax}{r^2} \cos \kappa \, (Vt \pm z), \quad Y = \frac{Ay}{r^2} \cos \kappa (Vt \pm z), \quad Z = 0,$$

where the axis of z is along the wire, $2\pi/\kappa$ is the wave length of the waves under consideration, A is a constant and $r^2 = x^2 + y^2$. The values of the components of the electrokinetic momentum are

$$F = -\frac{A}{\kappa V} \frac{x}{r^2} \sin \kappa \, (Vt \pm z), \quad G = -\frac{A}{\kappa V} \cdot \frac{y}{r^2} \sin \kappa \, (Vt \pm z), \quad H = 0,$$

and the components of the magnetic force are

$$\alpha = \pm \frac{A}{V} \frac{y}{r^2} \cos \kappa \, (Vt \pm z), \quad \beta = \mp \frac{A}{V} \frac{x}{r^2} \cos \kappa \, (Vt \pm z), \quad \gamma = 0.$$

The rate of radiation of energy, across the surface of a circular cylinder having the wire as axis, is

$$\frac{1}{8\pi} \frac{d}{dt} \iint [F(m\gamma - n\beta) + G(n\alpha - l\gamma) + H(l\beta - m\alpha)] \, dS$$

$$+ \frac{1}{4\pi} \iint [X(m\gamma - n\beta) + Y(n\alpha - l\gamma) + Z(l\beta - m\alpha)] \, dS,$$

where $\qquad\qquad l = \frac{x}{r}, \quad m = \frac{y}{r}, \quad n = 0.$

Now $\qquad\qquad \gamma = 0, \quad H = 0, \quad Z = 0,$

and also

$$m\gamma - n\beta = 0, \quad n\alpha - l\gamma = 0, \quad m\dot{\gamma} - n\dot{\beta} = 0, \quad n\dot{\alpha} - l\dot{\gamma} = 0 \, ;$$

therefore the above expression for the time rate of radiation of energy across the surface vanishes and, as was seen before, the waves are permanent.

The density of the distribution of the energy of the waves at any point is

$$\tfrac{1}{2}\,(F\dot{f}+G\dot{g}+H\dot{h}+Xf+Yg+Zh)$$

per unit volume, which is equivalent to

$$\frac{A^{2}}{8\pi V^{2}}\left[\frac{x^{2}+y^{2}}{r^{4}}\sin^{2}\kappa\,(Vt\pm z)+\frac{x^{2}+y^{2}}{r^{4}}\cos^{2}\kappa\,(Vt\pm z)\right],$$

that is to $A^{2}/8\pi V^{2}r^{2}$, so that it varies inversely as the square of the distance of the point from the wire.

It follows immediately that, for any number of parallel wires along which waves are travelling, the rate of radiation across any parallel cylindrical surface vanishes.

56. Taking now the case of waves travelling along any circuit, let the tubular surface generated by a sphere of small radius ρ, whose centre moves along the circuit, be considered. The electric force at a point on this surface, due to the waves in the circuit, consists of a part, which varies as $1/\rho$ and whose direction is along the normal to the surface, together with a part, which involves ρ as a factor since it must vanish with ρ^{*}; the same is true of the electrokinetic momentum. The magnetic force consists of a part, varying as $1/\rho$, in the plane through the point perpendicular to the circuit and tangential to the tubular surface, the remaining part not being infinite when ρ vanishes. Through two adjacent points P and Q of the circuit let planes perpendicular to PQ be drawn determining a strip on the tubular surface. Choosing PQ as axis of Z, the components of the electric force, at a point on the strip, are given by

$$X=\frac{x}{\rho^{2}}A+\rho X_{1},\quad Y=\frac{y}{\rho^{2}}A+\rho Y_{1},\quad Z=\rho Z_{1},$$

the components of the electrokinetic momentum by

$$F=\frac{x}{\rho^{2}}B+\rho F_{1},\quad G=\frac{y}{\rho^{2}}B+\rho G_{1},\quad H=\rho H_{1},$$

and the components of the magnetic force by

$$\alpha=\frac{y}{\rho^{2}}C+\alpha_{1},\quad \beta=-\frac{x}{\rho^{2}}C+\beta_{1},\quad \gamma=\gamma_{1},$$

---

* This follows from the expressions for the electric force, § 44.

6—2

where X_1, Y_1, Z_1, F_1, G_1, H_1, α_1, β_1, γ_1 do not become infinite when ρ vanishes. Then

$$F(m\gamma - n\beta) + G(n\alpha - l\gamma) + H(l\beta - m\alpha)$$
$$= \gamma_1(yF_1 - xG_1) + H_1(x\beta_1 - y\alpha_1),$$

and

$$X(m\gamma - n\beta) + Y(n\alpha - l\gamma) + Z(l\beta - m\alpha)$$
$$= \gamma_1(yX_1 - xY_1) + Z_1(x\beta_1 - y\alpha_1);$$

hence the rate of radiation of energy across the strip is $\rho^2\chi$, where χ does not become infinite when ρ vanishes, and therefore, for the whole surface, the rate of radiation of energy across it is $\rho^2\psi$, where ψ does not become infinite when ρ vanishes. Now (§ 47) the rate of radiation of energy across all surfaces, which enclose all the sources of the waves, is the same, and therefore in the above ψ vanishes, as otherwise the rate of radiation of energy, across the surface, would depend on its size; it therefore follows that the rate of radiation of energy across the tubular surface vanishes, and the waves are permanent as before.

CHAPTER IX.

OPEN CIRCUITS.

57. IT appears from the investigation of Chapter VII. that, to excite the waves of simple character discussed in § 44, some means must be found of enabling closed lines of magnetic force to thread the circuit. This can only be effected by cutting the circuit, so that these closed lines of magnetic force can pass over the open ends. It thus becomes necessary to investigate the effect of an open end on the propagation of waves along a circuit, and only those waves, for which the function denoted by L in § 44 has a finite value, need be discussed, as the others, in the case of a very thin wire, have been shewn to be comparatively unimportant and, in the limit, they have no existence. The waves, in the case of an open end, must be maintained by some external means, and the function L, at any point of the wire, is to be determined by the condition that $\dfrac{\partial^2 L}{\partial s^2} + \kappa^2 L$ is proportional to the electric force, tangential to the wire, due to the external disturbances, and L vanishes at the open end, this being necessary on account of the manner in which the expression for the tangential electric force was derived in § 44. When the function L has been determined, the radiation of the energy of the waves from the open end, from which only radiation of energy takes place by the previous investigations, can be obtained, so that, conversely, if the nature of the radiation from the open end were known, the function L could be obtained and the necessary sources; further, for any circuit having a given open end from which radiation is taking place in a given manner, the function L will be the same at the same distance along the circuit from the open end, provided that the curvature of the circuit is everywhere continuous. The same result will hold for a very thin wire, provided that the radius of its cross-section is everywhere small compared with the wave length and with the radius of curvature of the wire. It is therefore sufficient, when the circumstances of the radiation from the open end are known, to investigate the case where the circuit is a semi-infinite straight line. To solve the problem of radiation from a semi-infinite straight line, it is convenient to solve the problem for the case of a right circular cone, and deduce that of the semi-infinite straight line as the limit, when the vertical angle of the cone becomes indefinitely small.

58. The space to be considered is that outside the right circular cone which is determined by $\theta = \theta_0$, where θ is the angle made by a vector drawn from the vertex of the cone with a fixed direction, and the space occupied by the cone is that for which θ lies between θ_0 and π; then, choosing polar coordinates of which θ is one, and remembering that, in view of the preceding, the case required is that in which the lines of magnetic force are circles having their centres on the axis of the cone, the equations in the notation of § 36 are

$$4\pi u = \frac{1}{r^2 \sin \theta} \frac{\partial}{\partial \theta} (\gamma r \sin \theta),$$

$$4\pi v = - \frac{1}{r \sin \theta} \frac{\partial}{\partial r} (\gamma r \sin \theta),$$

$$4\pi w = 0,$$

$$-\frac{\partial \gamma}{\partial t} = \frac{1}{r} \left[\frac{\partial}{\partial r} (Yr) - \frac{\partial X}{\partial \theta} \right],$$

the medium being assumed to be non-magnetic. Hence, using the relations

$$(X, Y, Z) = 4\pi V^2 (f, g, h),$$

$$(u, v, w) = (\dot{f}, \dot{g}, \dot{h}),$$

the magnetic force γ satisfies the equation

$$\frac{1}{r} \frac{\partial}{\partial r} \left\{ \frac{1}{\sin \theta} \frac{\partial}{\partial r} (\gamma r \sin \theta) \right\} + \frac{1}{r^2} \frac{\partial}{\partial \theta} \left\{ \frac{1}{\sin \theta} \frac{\partial}{\partial \theta} (\gamma r \sin \theta) \right\} = \frac{1}{V^2} \frac{\partial^2 \gamma}{\partial t^2},$$

which, writing $\cos \theta = \mu$ and $\gamma r \sin \theta = \psi$, is equivalent to

$$\frac{\partial^2 \psi}{\partial r^2} + \frac{1 - \mu^2}{r^2} \frac{\partial^2 \psi}{\partial \mu^2} = \frac{1}{V^2} \frac{\partial^2 \psi}{\partial t^2}.$$

Therefore, for oscillations of definite wave length $2\pi/\kappa$, the equation satisfied by ψ is

$$\frac{\partial^2 \psi}{\partial r^2} + \frac{1 - \mu^2}{r^2} \frac{\partial^2 \psi}{\partial \mu^2} + \kappa^2 \psi = 0 \quad \dots\dots\dots\dots(1),$$

at all points at which the relations

$$(u, v, w) = (\dot{f}, \dot{g}, \dot{h})$$

are satisfied, that is at all points at which there are no sources. For the case under consideration the sources are distributed symmetrically with respect to the axis, and the discontinuities of the electric force are therefore symmetrically distributed with respect to the axis, but, instead of supposing the sources represented by discontinuities of the electric force, it is more convenient in this case to suppose them represented by discontinuities of the magnetic force as in § 14. The equation to be satisfied by the magnetic force at all points of a circle, whose centre is on the axis of the cone, and along which sources of this type of equal strength are uniformly distributed, is

$$\frac{\partial^2 \psi}{\partial r^2} + \frac{1 - \mu^2}{r^2} \frac{\partial^2 \psi}{\partial \mu^2} + \kappa^2 \psi + 2\omega r \sin \theta = 0 \dots\dots\dots(2),$$

where ω is a constant. The surface of the cone being taken to be perfectly conducting, the condition to be satisfied at its surface is that the tangential components of the electric force vanish, and this requires that

$$\frac{1}{r^2} \frac{\partial \psi}{\partial \mu} = 0 \dots\dots\dots\dots\dots\dots(3),$$

when $\theta = \theta_0$.

59. The solution of equation (2) is given by

$$\psi = \Sigma R_n (1 - \mu^2) \frac{\partial P_n}{\partial \mu},$$

where P_n is the zonal harmonic of μ of order n, and where the summation has to be taken so as to include all the functions of μ whose inclusion is consistent with the condition that $\dfrac{\partial \psi}{\partial \mu}$ vanishes, when $\theta = \theta_0$. Now

$$\frac{\partial}{\partial \mu}(1 - \mu^2)\frac{\partial P_n}{\partial \mu} = -n(n+1)P_n,$$

and therefore the functions P_n, which can occur, are those which vanish when $\mu = \mu_0$, that is, those for which n is a zero of $P_n(\mu_0)$, where $\mu_0 = \cos \theta_0$. Hence

$$\psi = \Sigma R_n (1 - \mu^2)\frac{\partial P_n}{\partial \mu},$$

where the summation extends to all the values of n which are zeros of $P_n(\mu_0)$; these values are all real. The functions R_n are then to be determined from the relation

$$\Sigma \left\{ \frac{\partial^2 R_n}{\partial r^2}(1 - \mu^2)\frac{\partial P_n}{\partial \mu} + R_n \frac{1 - \mu^2}{r^2}\frac{\partial^2}{\partial \mu^2} \cdot (1 - \mu^2)\frac{\partial P_n}{\partial \mu} \right.$$
$$\left. + \kappa^2 R_n (1 - \mu^2)\frac{\partial P_n}{\partial \mu} \right\} + 2\omega r \sin \theta = 0,$$

which is equivalent to

$$\Sigma \left[\frac{\partial^2 R_n}{\partial r^2} + \left\{ \kappa^2 - \frac{n(n+1)}{r^2} \right\} R_n \right] (1 - \mu^2)\frac{\partial P_n}{\partial \mu} + 2\omega r \sin \theta = 0 \dots (4).$$

Now

$$n(n+1)\int_{\mu_0}^1 (1 - \mu^2)\frac{\partial P_n}{\partial \mu}\frac{\partial P_{n'}}{\partial \mu}\,d\mu$$
$$= -\int_{\mu_0}^1 (1 - \mu^2)\frac{\partial P_{n'}}{\partial \mu}\frac{\partial^2}{\partial \mu^2}\left\{ (1 - \mu^2)\frac{\partial P_n}{\partial \mu} \right\} d\mu$$

and therefore, writing

$$y = (1 - \mu^2)\frac{\partial P_n}{\partial \mu}, \quad y' = (1 - \mu^2)\frac{\partial P_{n'}}{\partial \mu},$$

$$\{ n'(n'+1) - n(n+1) \}\int_{\mu_0}^1 (1 - \mu^2)\frac{\partial P_n}{\partial \mu}\frac{\partial P_{n'}}{\partial \mu}\,d\mu$$
$$= \int_{\mu_0}^1 \left\{ y'\frac{d^2 y}{d\mu^2} - y\frac{d^2 y'}{d\mu^2} \right\} d\mu,$$

that is

$$(n' - n)(n' + n + 1)\int_{\mu_0}^1 (1 - \mu^2)\frac{\partial P_n}{\partial \mu}\frac{\partial P_{n'}}{\partial \mu}\,d\mu = \left| y'\frac{dy}{d\mu} - y\frac{dy'}{d\mu} \right|_{\mu_0}^1.$$

Hence, writing for y and y' their values and reducing,

$$(n' - n)(n' + n + 1) \int_{\mu_0}^{1} (1 - \mu^2) \frac{\partial P_n}{\partial \mu} \frac{\partial P_{n'}}{\partial \mu} \, d\mu$$

$$= (1 - \mu_0^2) \left[n(n+1) P_n \frac{\partial P_{n'}}{\partial \mu} - n'(n'+1) P_{n'} \frac{\partial P_n}{\partial \mu} \right],$$

where, on the right-hand side, μ_0 is to be written for μ after the operations have been performed. From this it follows that, if n and n' are two positive values of n for which $P_n(\mu_0)$ vanishes, it being remembered that only positive values need be considered as

$$P_n(\mu) = P_{-n-1}(\mu),$$

the integral

$$\int_{\mu_0}^{1} (1 - \mu^2) \frac{\partial P_n}{\partial \mu} \frac{\partial P_{n'}}{\partial \mu} \, d\mu$$

vanishes. To obtain the value of this integral when n and n' are the same, let

$$n' = n + \delta n,$$

then the relation given above becomes

$$(2n + 1) \delta n \int_{\mu_0}^{1} (1 - \mu^2) \frac{\partial P_n}{\partial \mu} \frac{\partial P_n}{\partial \mu} \, d\mu$$

$$= - n(n+1)(1 - \mu_0^2) \frac{\partial P_n}{\partial n} \frac{\partial P_n}{\partial \mu} \delta n,$$

and therefore

$$\int_{\mu_0}^{1} (1 - \mu^2) \frac{\partial P_n}{\partial \mu} \frac{\partial P_n}{\partial \mu} = - \frac{n(n+1)}{2n+1} (1 - \mu_0^2) \frac{\partial P_n}{\partial n} \frac{\partial P_n}{\partial \mu},$$

where, on the right-hand side, μ_0 is to be written for μ after the operations have been performed. Multiplying both sides of the equation (4) by $\frac{\partial P_n}{\partial \mu} \, d\mu$ and integrating with respect to μ from μ_0 to 1, the equation to determine R_n is

$$\left[\frac{\partial^2 R_n}{\partial r^2} + \left\{ \kappa^2 - \frac{n(n+1)}{r^2} \right\} R_n \right] \frac{n(n+1)}{2n+1} (1 - \mu_0^2) \frac{\partial P_n}{\partial n} \frac{\partial P_n}{\partial \mu_0}$$

$$= 2r \int_{\mu_0}^{1} \omega \sin \theta \frac{\partial P_n}{\partial \mu} \, d\mu \dots\dots(5).$$

It is sufficient to consider the case where the sources are distributed uniformly upon the circumference of the circle determined by the relations

$$r = r_1, \quad \theta = \theta_1,$$

θ_1 being less than θ_0; any other case can be obtained from this by a process of summation. In this case the solution of equation (5) will have different forms according as r is greater or less than r_1. It being remembered that the right-hand side of (5) vanishes unless $r = r_1$, the solution when r is less than r_1 is given in terms of Bessel's functions by

$$R_n = A_n r^{\frac{1}{2}} J_{n+\frac{1}{2}}(\kappa r),$$

as R_n cannot be infinite at the vertex of the cone, that is when $r = 0$; also when r is greater than r_1 the solution is given by

$$R_n = B_n r^{\frac{1}{2}} \{ J_{-n-\frac{1}{2}}(\kappa r) - e^{(n+\frac{1}{2})\pi \iota} J_{n+\frac{1}{2}}(\kappa r) \},$$

as at an infinite distance R_n cannot involve a term in which $e^{\iota \kappa r}$ occurs, there being no reflexion there. Now, when these expressions are substituted in the expression for ψ, two series are obtained for ψ, one of which is applicable when r is less than r_1, and the other when r is greater than r_1, these series converging for all values of r and θ which occur, except when $r = r_1$ and $\theta = \theta_1$ simultaneously. Therefore, when $r = r_1$, the two series must be identical; whence

$$A_n J_{n+\frac{1}{2}}(\kappa r_1) = B_n \{ J_{-n-\frac{1}{2}}(\kappa r_1) - e^{(n+\frac{1}{2})\pi \iota} J_{n+\frac{1}{2}}(\kappa r_1) \},$$

and the solution of (5) may be written

$$R_n = C_n r^{\frac{1}{2}} J_{n+\frac{1}{2}}(\kappa r) \{ J_{-n-\frac{1}{2}}(\kappa r_1) - e^{(n+\frac{1}{2})\pi \iota} J_{n+\frac{1}{2}}(\kappa r_1) \},$$

when $\qquad\qquad r < r_1$ ....(6),

$$R_n = C_n r^{\frac{1}{2}} J_{n+\frac{1}{2}}(\kappa r_1) \{ J_{-n-\frac{1}{2}}(\kappa r) - e^{(n+\frac{1}{2})\pi \iota} J_{n+\frac{1}{2}}(\kappa r) \},$$

when $\qquad\qquad r > r_1$ ....(6').

60. It now remains to determine C_n. The direct determination of C_n, though possible, involves somewhat complicated analysis, which can be evaded by the observation that C_n is independent of the wave length, that is of κ. To establish this, it is sufficient to prove it for a particular case, which can be chosen to be that in which there is no cone and only the circle of sources radiating freely into space. In this case the values of n which occur are all the positive integers. It can be easily shewn that equation (2) is equivalent to the equation

$$(\nabla^2 + \kappa^2) \frac{\psi \sin \phi}{r \sin \theta} + 2\omega \sin \phi = 0 \quad \ldots\ldots\ldots\ldots(2'),$$

where ϕ is the third polar coordinate, viz. the longitude. The solution of equation $(2')$ in the particular case, where there is no cone, is given by

$$\frac{\psi \sin \phi}{r \sin \theta} = \frac{1}{2\pi} \iiint \frac{e^{-\iota \kappa R}}{R} \, \omega r_1^2 \sin \theta_1 \sin \phi_1 d\theta_1 d\phi_1 dr_1,$$

where the integration is taken so as to include all places where ω has a value different from zero, and

$$R^2 = r^2 + r_1^2 - 2rr_1 \{\cos \theta \cos \theta_1 + \sin \theta \sin \theta_1 \cos (\phi - \phi_1)\}.$$

Now, writing

$$\cos \gamma = \cos \theta \cos \theta_1 + \sin \theta \sin \theta_1 \cos (\phi - \phi_1),$$

the integrand is expressed as a series involving spherical harmonics by means of the relations*

$$\frac{e^{-\iota \kappa R}}{R} = \frac{2e^{\frac{\pi \iota}{4}}}{\sqrt{(rr_1)}} \sum_0^\infty e^{\frac{n \pi \iota}{2}} (n + \tfrac{1}{2}) \, K_{n+\frac{1}{2}} (\iota \kappa r_1) \, J_{n+\frac{1}{2}} (\kappa r) \, P_n (\cos \gamma),$$

when $r < r_1$ and

$$\frac{e^{-\iota \kappa R}}{R} = \frac{2e^{\frac{\pi \iota}{4}}}{\sqrt{(rr_1)}} \sum_0^\infty e^{\frac{n \pi \iota}{2}} (n + \tfrac{1}{2}) \, K_{n+\frac{1}{2}} (\iota \kappa r) \, J_{n+\frac{1}{2}} (\kappa r_1) \, P_n (\cos \gamma),$$

when $r > r_1$; whence

$$\frac{\psi \sin \phi}{r \sin \theta} = \frac{e^{\frac{\pi \iota}{4}}}{\pi} \iiint \omega r_1^2 \sin \theta_1 \sin \phi_1 \, d\theta_1 d\phi_1 dr_1 \frac{1}{\sqrt{(rr_1)}} \sum_0^\infty e^{\frac{n \pi \iota}{2}} (n + \tfrac{1}{2})$$
$$. K_{n+\frac{1}{2}} (\iota \kappa r_1) \, J_{n+\frac{1}{2}} (\kappa r) \, P_n (\cos \gamma),$$

when $r < r_1$, and for a circle of sources, writing

$$q = \omega r_1^2 \sin \theta_1 d\theta_1 dr_1,$$

$$\psi \sin \phi = \frac{qr \sin \theta \, e^{\frac{\pi \iota}{4}}}{\pi \sqrt{(rr_1)}} \sum_0^\infty e^{\frac{n \pi \iota}{2}} (n + \tfrac{1}{2}) \, K_{n+\frac{1}{2}} (\iota \kappa r_1)$$
$$. J_{n+\frac{1}{2}} (\kappa r) \int_0^{2\pi} P_n (\cos \gamma) \sin \phi_1 d\phi_1.$$

---

* Heine, *Handbuch der Kugelfunctionen*, I. p. 346; Macdonald, *Proc. Lond. Math. Soc.* Vol. XXXII. 1900.

Therefore, since

$$\int_0^{2\pi} P_n (\cos \gamma) \sin \phi_1 d\phi_1 = \frac{2\pi}{n(n+1)} \sin \theta \sin \theta_1 \frac{\partial P_n}{\partial \mu} \frac{\partial P_n}{\partial \mu_1} \sin \phi,$$

it follows that

$$\psi = qr^{\frac{1}{2}}r_1^{-\frac{1}{2}} e^{\frac{\pi \iota}{4}} \sum_0^\infty e^{\frac{n\pi \iota}{2}} \frac{2n+1}{n(n+1)} \sin^2 \theta \sin \theta_1 \frac{\partial P_n}{\partial \mu} \frac{\partial P_n}{\partial \mu_1}$$
$$. J_{n+\frac{1}{2}} (\kappa r) K_{n+\frac{1}{2}} (\iota \kappa r_1),$$

when $\qquad\qquad\qquad r < r_1$ ....(7),

and

$$\psi = qr^{\frac{1}{2}}r_1^{-\frac{1}{2}} e^{\frac{\pi \iota}{4}} \sum_0^\infty e^{\frac{n\pi \iota}{2}} \frac{2n+1}{n(n+1)} \sin^2 \theta \sin \theta_1 \frac{\partial P_n}{\partial \mu} \frac{\partial P_n}{\partial \mu_1}$$
$$. J_{n+\frac{1}{2}} (\kappa r_1) K_{n+\frac{1}{2}} (\iota \kappa r),$$

when $\qquad\qquad\qquad r > r_1$ ....(7').

Remembering that

$$K_m (\iota \kappa r) = \frac{\pi e^{-\frac{m\pi \iota}{2}}}{2 \sin m\pi} \{ J_{-m}(\kappa r) - e^{m\pi \iota} J_m (\kappa r) \},$$

and comparing with the solutions obtained by substituting the expressions (6) and (6') for R_n in

$$\psi = \Sigma R_n (1 - \mu^2) \frac{\partial P_n}{\partial \mu},$$

it follows that C_n is independent of κ, that is of the wave length.

61. When κ is zero, the equation (5) becomes

$$\left[\frac{\partial^2 R_n}{\partial r^2} - \frac{n(n+1)}{r^2} R_n \right] \frac{n(n+1)}{2n+1} (1 - \mu_0^2) \frac{\partial P_n}{\partial n} \frac{\partial P_n}{\partial \mu_0}$$
$$= 2r \int_{\mu_0}^1 \omega \sin \theta \frac{\partial P_n}{\partial \mu} d\mu,$$

which, putting

$$R_n = r^{\frac{1}{2}} L_n,$$

and writing

$$\frac{n(n+1)}{2n+1} (1 - \mu_0^2) \frac{\partial P_n}{\partial n} \frac{\partial P_n}{\partial \mu_0} = N,$$

becomes

$$\frac{\partial^2 L_n}{\partial r^2} + \frac{1}{r} \frac{\partial L_n}{\partial r} - \frac{(n+\frac{1}{2})^2}{r^2} L_n = \frac{2r^{\frac{1}{2}}}{N} \int_{\mu_0}^1 \omega \sin \theta \frac{\partial P_n}{\partial \mu} d\mu,$$

that is, making the substitution

$$r = ae^{-\xi},$$

$$\frac{\partial^2 L_n}{\partial \xi^2} - (n + \tfrac{1}{2})^2 L_n = \frac{2a^{\frac{5}{2}} e^{-\frac{5\xi}{2}}}{N} \int_{\mu_0}^1 \omega \sin \theta \frac{\partial P_n}{\partial \mu} d\mu.$$

It is convenient to solve this equation for the range of ξ from zero to ξ_0 and then to proceed to the limit; the conditions, corresponding to those to be satisfied in the limiting case, are

$$L_n = 0, \text{ when } \xi = 0 \text{ and when } \xi = \xi_0,$$

hence it may be assumed that

$$L_n = \sum_1^\infty A_m \sin \frac{m\pi\xi}{\xi_0},$$

where m is an integer. Substituting in the equation for L_n, it becomes

$$-\sum_1^\infty A_m \left\{ \frac{m^2 \pi^2}{\xi_0^2} + (n + \tfrac{1}{2})^2 \right\} \sin \frac{m\pi\xi}{\xi_0} = \frac{2a^{\frac{5}{2}} e^{-\frac{5\xi}{2}}}{N} \int_{\mu_0}^1 \omega \sin \theta \frac{\partial P_n}{\partial \mu} d\mu,$$

and therefore

$$-A_m \left\{ \frac{m^2 \pi^2}{\xi_0^2} + (n + \tfrac{1}{2})^2 \right\} = \frac{4a^{\frac{5}{2}}}{N\xi_0} \int_0^{\xi_0} e^{-\frac{5\xi}{2}} \sin \frac{m\pi\xi}{\xi_0} d\xi \int_{\mu_0}^1 \omega \sin \theta \frac{\partial P_n}{\partial \mu} d\mu.$$

For a single circle of sources, writing

$$q = \omega a^3 e^{-3\xi_1} \sin \theta_1 d\theta_1 d\xi_1,$$

the circle being defined by

$$\theta = \theta_1, \quad \xi = \xi_1,$$

this becomes

$$A_m \left\{ \frac{m^2 \pi^2}{\xi_0^2} + (n + \tfrac{1}{2})^2 \right\} = \frac{4qa^{-\frac{1}{2}} e^{\frac{\xi_1}{2}}}{N\xi_0} \sin \theta_1 \frac{\partial P_n}{\partial \mu_1} \sin \frac{m\pi\xi_1}{\xi_0},$$

and therefore

$$L_n = \sum_1^\infty \frac{4qa^{-\frac{1}{2}} e^{\frac{\xi_1}{2}}}{N\xi_0} \sin \theta_1 \frac{\partial P_n}{\partial \mu_1} \cdot \frac{\sin \dfrac{m\pi\xi}{\xi_0} \sin \dfrac{m\pi\xi_1}{\xi_0}}{\dfrac{m^2 \pi^2}{\xi_0^2} + (n + \tfrac{1}{2})^2}.$$

Effecting the summation, L_n is given by

$$L_n = - \frac{2qa^{-\frac{1}{2}}e^{\frac{\xi_1}{2}}}{(2n+1)N} \sin \theta_1 \frac{\partial P_n}{\partial \mu_1}$$
$$\cdot \frac{\cosh (n+\tfrac{1}{2})(\xi_0 - \xi + \xi_1) - \cosh (n+\tfrac{1}{2})(\xi_0 - \xi - \xi_1)}{\sinh (n+\tfrac{1}{2}) \xi_0},$$

when $\xi > \xi_1$ and by

$$L_n = - \frac{2qa^{-\frac{1}{2}}e^{\frac{\xi_1}{2}}}{(2n+1)N} \sin \theta_1 \frac{\partial P_n}{\partial \mu_1}$$
$$\cdot \frac{\cosh (n+\tfrac{1}{2})(\xi_0 - \xi_1 + \xi) - \cosh (n+\tfrac{1}{2})(\xi_0 - \xi - \xi_1)}{\sinh (n+\tfrac{1}{2}) \xi_0},$$

when $\xi < \xi_1$. Making ξ_0 infinite, the expression for L_n becomes

$$L_n = - \frac{2qa^{-\frac{1}{2}}e^{\frac{\xi_1}{2}}}{(2n+1)N} \sin \theta_1 \frac{\partial P_n}{\partial \mu_1} \{ e^{-(n+\frac{1}{2})(\xi - \xi_1)} - e^{-(n+\frac{1}{2})(\xi + \xi_1)} \},$$

when $\xi > \xi_1$, and

$$L_n = - \frac{2qa^{-\frac{1}{2}}e^{\frac{\xi_1}{2}}}{(2n+1)N} \sin \theta_1 \frac{\partial P_n}{\partial \mu_1} \{ e^{-(n+\frac{1}{2})(\xi_1 - \xi)} - e^{-(n+\frac{1}{2})(\xi + \xi_1)} \},$$

when $\xi < \xi_1$, that is

$$L_n = - \frac{2qr_1^{-\frac{1}{2}}}{(2n+1)N} \left\{ \left(\frac{r}{r_1} \right)^{n+\frac{1}{2}} - \left(\frac{rr_1}{a^2} \right)^{n+\frac{1}{2}} \right\} \sin \theta_1 \frac{\partial P_n}{\partial \mu_1},$$

when $r < r_1$, and

$$L_n = - \frac{2qr_1^{-\frac{1}{2}}}{(2n+1)N} \left\{ \left(\frac{r_1}{r} \right)^{n+\frac{1}{2}} - \left(\frac{rr_1}{a^2} \right)^{n+\frac{1}{2}} \right\} \sin \theta_1 \frac{\partial P_n}{\partial \mu_1},$$

when $r > r_1$. In the case required, the range of r is from 0 to ∞, and this case is obtained from the above by making a infinite; hence

$$L_n = - \frac{2qr_1^{-\frac{1}{2}}}{(2n+1)N} \left(\frac{r}{r_1} \right)^{n+\frac{1}{2}} \sin \theta_1 \frac{\partial P_n}{\partial \mu_1},$$

when $r < r_1$, and

$$L_n = - \frac{2qr_1^{-\frac{1}{2}}}{(2n+1)N} \left(\frac{r_1}{r} \right)^{n+\frac{1}{2}} \sin \theta_1 \frac{\partial P_n}{\partial \mu_1},$$

when $r > r_1$. Now, when $\kappa = 0$, equation (6) becomes

$$R_n = C_n r^{\frac{1}{2}} \frac{r^{n+\frac{1}{2}}}{2^{n+\frac{1}{2}} \Pi (n + \frac{1}{2})} \cdot \frac{r_1^{-n-\frac{1}{2}}}{2^{-n-\frac{1}{2}} \Pi (-n - \frac{1}{2})},$$

when $r < r_1$, that is, since

$$R_n = r^{\frac{1}{2}} L_n,$$

$$L_n = C_n \frac{\sin (n + \frac{1}{2}) \pi}{(n + \frac{1}{2}) \pi} \left(\frac{r}{r_1}\right)^{n+\frac{1}{2}},$$

when $r < r_1$, and similarly from equation (6′)

$$L_n = C_n \frac{\sin (n + \frac{1}{2}) \pi}{(n + \frac{1}{2}) \pi} \left(\frac{r_1}{r}\right)^{n+\frac{1}{2}},$$

when $r > r_1$; therefore

$$C_n = - \frac{q r_1^{-\frac{1}{2}}}{N} \cdot \frac{\pi}{\sin (n + \frac{1}{2}) \pi} \sin \theta_1 \frac{\partial P_n}{\partial \mu_1}.$$

62. The solution, for a circle of sources emitting waves of wave length $2\pi/\kappa$, the circle being defined by $r = r_1$, $\theta = \theta_1$ and being situated in the space external to the cone defined by $\theta = \theta_0$, is therefore given by

$$\psi = - q \Sigma \left(\frac{r}{r_1}\right)^{\frac{1}{2}} J_{n+\frac{1}{2}} (\kappa r) \{J_{-n-\frac{1}{2}} (\kappa r_1) - e^{(n+\frac{1}{2})\pi\iota} J_{n+\frac{1}{2}} (\kappa r_1)\}$$

$$\cdot \frac{\pi}{N \sin (n + \frac{1}{2}) \pi} \sin \theta_1 \frac{\partial P_n}{\partial \mu_1} (1 - \mu^2) \frac{\partial P_n}{\partial \mu},$$

when $\qquad\qquad r < r_1$ ....(8),

and by

$$\psi = - q \Sigma \left(\frac{r}{r_1}\right)^{\frac{1}{2}} J_{n+\frac{1}{2}} (\kappa r_1) \{J_{-n-\frac{1}{2}} (\kappa r) - e^{(n+\frac{1}{2})\pi\iota} J_{n+\frac{1}{2}} (\kappa r)\}$$

$$\cdot \frac{\pi}{N \sin (n + \frac{1}{2}) \pi} \sin \theta_1 \frac{\partial P_n}{\partial \mu_1} (1 - \mu^2) \frac{\partial P_n}{\partial \mu},$$

when $\qquad\qquad r > r_1$ ....(8′),

where the time-factor is included in q, N is written for

$$\frac{n (n + 1)}{2n + 1} (1 - \mu_0^2) \frac{\partial P_n}{\partial n} \frac{\partial P_n}{\partial \mu_0},$$

and the summation is extended to all the positive values of n which make $P_n (\mu_0)$ vanish.

In the particular case where $\theta_0 = \dfrac{\pi}{2}$, which is the case of a circle of sources in the space bounded by an infinite conducting plane, the values of n which occur are all the positive odd integers, the value of $\dfrac{\partial P_n}{\partial n}\dfrac{\partial P_n}{\partial \mu_0}$ is -1 and, remembering equations (7) and (7′) above, the expression for ψ is seen to be the sum of the ψ due to the circle of sources and the ψ due to the image of this circle of sources with respect to the plane, a result which is otherwise immediately obvious.

63. The case of an indefinitely thin semi-infinite straight wire can be obtained from the above by taking $\pi - \epsilon$ as the value of θ_0, where ϵ is very small. The values of n which occur are then given by

$$n = k + n_0$$

where

$$2n_0 \log \frac{2}{\epsilon} = 1,$$

and k may be any positive integer or zero. Further, since

$$P_n(-\mu) = \cos n\pi \, P_n(\mu) - \frac{2}{\pi} \sin n\pi \, Q_n(\mu),$$

it follows that

$$\frac{\partial P_n(-\mu)}{\partial \mu} = \cos n\pi \, \frac{\partial P_n(\mu)}{\partial \mu} - \frac{2}{\pi} \sin n\pi \, \frac{\partial Q_n(\mu)}{\partial \mu};$$

that is, writing $\mu_0 = -\mu_0'$,

$$\frac{\partial P_n(\mu_0)}{\partial \mu_0} = -\cos n\pi \, \frac{\partial P_n(\mu_0')}{\partial \mu_0'} + \frac{2}{\pi} \sin n\pi \, \frac{\partial Q_n(\mu_0')}{\partial \mu_0'}.$$

Again*

$$Q_n(\mu) = \tfrac{1}{2} P_n(\mu) \log \frac{1+\mu}{1-\mu} + \left\{ \Pi'(0) - \frac{\Pi'(n)}{\Pi(n)} \right\} P_n(\mu)$$

$$- \sum_1^\infty \frac{\Pi(-n+r-1)\,\Pi(n+r)}{\Pi(n)\,\Pi(-n-1)\,\Pi(r)\,\Pi(r)} \left(\frac{1-\mu}{2} \right)^r A_r,$$

where

$$A_r = \sum_1^r \frac{1}{m},$$

* Macdonald, *Proc. Lond. Math. Soc.* Vol. **xxxi.** 1899.

hence, omitting terms of the first order of small quantities,

$$(1 - \mu_0{}^2)\frac{\partial Q_n}{\partial \mu_0'} = 1,$$

and, observing that $(1 - \mu_0{}^2)/n_0$ ultimately vanishes,

$$(1 - \mu_0{}^2)\frac{\partial P_n}{\partial \mu_0} = \frac{2}{\pi} \sin n\pi,$$

the most significant term only being retained.

From the relation between $P_n(\mu)$ and the harmonics of $-\mu$, it follows that

$$\left(\frac{\partial P_n}{\partial n}\right)_{\mu = \mu_0} = -\pi \sin n\pi \, P_n(\mu_0') - 2 \cos n\pi \, Q_n(\mu_0')$$
$$+ \cos n\pi \frac{\partial P_n(\mu_0')}{\partial n} - \frac{2}{\pi} \sin n\pi \frac{\partial Q_n(\mu_0')}{\partial n},$$

that is, since $P_n(\mu_0) = 0$,

$$\left(\frac{\partial P_n}{\partial n}\right)_{\mu = \mu_0} = -\frac{\pi}{\sin n\pi} P_n(\mu_0') + \cos n\pi \frac{\partial P_n(\mu_0')}{\partial n}$$
$$- \frac{2}{\pi} \sin n\pi \frac{\partial Q_n(\mu_0')}{\partial \mu_0'},$$

which, all but the most important part being rejected, becomes

$$\left(\frac{\partial P_n}{\partial n}\right)_{\mu = \mu_0} = -\frac{\pi}{\sin n\pi}.$$

The value of N, omitting small terms, therefore is

$$-\frac{2k(k+1)}{2k+1},$$

if n has any of its values other than n_0, and it is $-2n_0$, if n is n_0.

When μ differs from μ_0 by a finite quantity and n has any of its values except n_0

$$(1 - \mu^2)\frac{\partial P_n}{\partial \mu} = (1 - \mu^2)\frac{\partial P_k}{\partial \mu},$$

neglecting small quantities, and when $n = n_0$

$$\frac{\partial P_{n_0}}{\partial \mu} = \frac{\partial}{\partial \mu} \sum_0^\infty \frac{\Pi(n_0 + s) \cos s\pi}{\Pi(n_0 - s)\, \Pi(s)\, \Pi(s)} \left(\frac{1 - \mu}{2}\right)^s,$$

that is

$$\frac{\partial P_{n_0}}{\partial \mu} = \frac{n_0}{2} \sum_1^\infty \left(\frac{1-\mu}{2}\right)^{s-1},$$

neglecting small quantities of the second order, hence, to this order,

$$\frac{\partial P_{n_0}}{\partial \mu} = \frac{n_0}{1+\mu}.$$

For the case of an indefinitely thin wire, the sources of the waves not being very close to the wire, that is θ_1 not being nearly equal to θ_0, substituting the above values in the equations (8) and (8′), ψ is given by

$$\psi = q \sum_1^\infty \left(\frac{r}{r_1}\right)^{\frac{1}{2}} J_{k+\frac{1}{2}}(\kappa r) \left\{J_{-k-\frac{1}{2}}(\kappa r_1) - e^{(k+\frac{1}{2})\pi i} J_{k+\frac{1}{2}}(\kappa r_1)\right\}$$

$$\cdot \frac{\pi}{\sin\left(k+\frac{1}{2}\right)\pi} \cdot \frac{2k+1}{2k(k+1)} \sin\theta_1 \frac{\partial P_k}{\partial \mu_1} (1-\mu^2) \frac{\partial P_k}{\partial \mu},$$

when $\qquad\qquad r < r_1$ ....(9),

and by

$$\psi = q \sum_1^\infty \left(\frac{r}{r_1}\right)^{\frac{1}{2}} J_{k+\frac{1}{2}}(\kappa r_1) \left\{J_{-k-\frac{1}{2}}(\kappa r) - e^{(k+\frac{1}{2})\pi i} J_{k+\frac{1}{2}}(\kappa r)\right\}$$

$$\cdot \frac{\pi}{\sin\left(k+\frac{1}{2}\right)\pi} \cdot \frac{2k+1}{2k(k+1)} \sin\theta_1 \frac{\partial P_k}{\partial \mu_1} (1-\mu^2) \frac{\partial P_k}{\partial \mu},$$

when $\qquad\qquad r > r_1$ ....(9′),

where μ is not nearly equal to μ_0. It thus appears, comparing this with equations (7) and (7′), that the effect of the indefinitely thin wire at points, whose distance from the free end of the wire is not very great compared with r_1, and for which μ is not nearly equal to μ_0, is negligible in comparison with the effect of the source. At a point, whose distance from the free end is great compared with r_1 and for which μ is not nearly equal to μ_0, the expression in (9′) has to have the term $n_0 e^{-\iota \kappa r} f(\mu)$ added to it, which represents the effect of the free end.

64. In the immediate neighbourhood of the wire μ differs but slightly from μ_0 and then

$$(1-\mu^2) \frac{\partial P_n}{\partial \mu} = \frac{2 \sin n\pi}{\pi},$$

that is

$$(1 - \mu^2)\frac{\partial P_n}{\partial \mu} = 2n_0 \cos k\pi,$$

and the first term of the series in (8) and (8′) is no longer negligible in comparison with the others. The value of the first term in (8) now is

$$q\left(\frac{r}{r_1}\right)^{\frac{1}{2}} J_{\frac{1}{2}}(\kappa r)\{J_{-\frac{1}{2}}(\kappa r_1) - e^{\frac{\pi \iota}{2}} J_{\frac{1}{2}}(\kappa r_1)\}$$

$$\cdot \frac{\pi}{2n_0}\frac{n_0^2}{(1 + \mu)(1 + \mu_1)}\sin \theta_1 (1 - \mu^2),$$

that is

$$\frac{2qn_0}{\kappa r_1}\sin \kappa r\, e^{-\iota \kappa r_1}\frac{\sin \theta_1}{1 + \mu_1},$$

and (8) becomes

$$\psi = \frac{2qn_0}{\kappa r_1}\sin \kappa r\, e^{-\iota \kappa r_1}\frac{\sin \theta_1}{1 + \mu_1} + 2qn_0\sum_1^{\infty}\left(\frac{r}{r_1}\right)^{\frac{1}{2}} J_{k+\frac{1}{2}}(\kappa r)$$

$$\cdot \{J_{-k-\frac{1}{2}}(\kappa r_1) - e^{(k+\frac{1}{2})\pi \iota} J_{k+\frac{1}{2}}(\kappa r_1)\}\frac{2k + 1}{2k(k + 1)}\sin \theta_1 \frac{\partial P_n}{\partial \mu_1},$$

when $r < r_1$, with a similar expression for ψ when $r > r_1$, giving the value of ψ along the wire. This series is capable of summation; writing

$$R^2 = r^2 + r_1^2 + 2rr_1 \cos \theta_1,$$

the expansion theorem for $\dfrac{e^{-\iota \kappa R}}{R}$ becomes*, remembering that

$$P_k(-\cos \theta_1) = (-)^k P_k(\cos \theta_1),$$

$$\frac{e^{-\iota \kappa R}}{R} = \frac{\pi}{r^{\frac{1}{2}}r_1^{\frac{1}{2}}}\sum_0^{\infty}(k + \tfrac{1}{2}) J_{k+\frac{1}{2}}(\kappa r)\{J_{-k-\frac{1}{2}}(\kappa r_1) - e^{(k+\frac{1}{2})\pi \iota} J_{k+\frac{1}{2}}(\kappa r_1)\}$$

$$\cdot P_k(\cos \theta_1),$$

when $r < r_1$, whence

$$\frac{e^{-\iota \kappa R}}{R} = \frac{\pi}{r^{\frac{1}{2}}r_1^{\frac{1}{2}}}\tfrac{1}{2} J_{\frac{1}{2}}(\kappa r)\{J_{-\frac{1}{2}}(\kappa r_1) - e^{\frac{\pi \iota}{2}} J_{\frac{1}{2}}(\kappa r_1)\}$$

$$- \frac{\pi}{r^{\frac{1}{2}}r_1^{\frac{1}{2}}}\sum_1^{\infty}\frac{2k + 1}{2k(k + 1)}J_{k+\frac{1}{2}}(\kappa r)\{J_{-k-\frac{1}{2}}(\kappa r_1) - e^{(k+\frac{1}{2})\pi \iota} J_{k+\frac{1}{2}}(\kappa r_1)\}$$

$$\cdot \frac{\partial}{\partial \mu_1}(1 - \mu_1^2)\frac{\partial P_k}{\partial \mu_1}.$$

* p. 91.

7—2

Therefore, writing

$$\chi = \pi \sum_{1}^{\infty} \frac{2k+1}{2k\,(k+1)} J_{k+\frac{1}{2}}(\kappa r)\left\{ J_{-k-\frac{1}{2}}(\kappa r_1) - e^{(k+\frac{1}{2})\pi \iota} J_{k+\frac{1}{2}}(\kappa r_1)\right\}$$
$$. \sin \theta_1 \frac{\partial P_k}{\partial \mu_1},$$

it follows that

$$\frac{e^{-\iota \kappa R}}{R} = \frac{e^{-\iota \kappa r_1}}{\kappa r r_1} \sin \kappa r - \frac{1}{r^{\frac{1}{2}} r_1^{\frac{1}{2}}} \frac{\partial}{\partial \mu_1}(\chi \sin \theta_1);$$

now
$$R \frac{\partial R}{\partial \mu_1} = r r_1,$$

hence

$$e^{-\iota \kappa R} \frac{\partial R}{\partial \mu_1} = \frac{e^{-\iota \kappa r_1}}{\kappa} \sin \kappa r - r^{\frac{1}{2}} r_1^{\frac{1}{2}} \frac{\partial}{\partial \mu_1}(\chi \sin \theta_1),$$

and therefore

$$\frac{\iota}{\kappa} e^{-\iota \kappa R} = \frac{\mu_1}{\kappa} e^{-\iota \kappa r_1} \sin \kappa r - r^{\frac{1}{2}} r_1^{\frac{1}{2}} \chi \sin \theta_1 + C,$$

where C is independent of θ_1. Substituting for θ_1 the value zero, it follows that, since χ is not infinite when θ_1 vanishes,

$$C = \frac{\iota}{\kappa} e^{-\iota \kappa (r+r_1)} - \frac{1}{\kappa} e^{-\iota \kappa r_1} \sin \kappa r,$$

that is

$$C = \frac{\iota}{\kappa} e^{-\iota \kappa r_1} \cos \kappa r;$$

and therefore

$$r^{\frac{1}{2}} r_1^{\frac{1}{2}} \chi \sin \theta_1 = \frac{1}{\kappa} e^{-\iota \kappa r_1}(\mu_1 \sin \kappa r + \iota \cos \kappa r) - \frac{\iota}{\kappa} e^{-\iota \kappa R}.$$

The value of ψ along the wire, when $r < r_1$, is given by

$$\psi = \frac{2q n_0}{\kappa r_1} e^{-\iota \kappa r_1} \sin \kappa r \frac{\sin \theta_1}{1 + \mu_1} + 2q n_0 \left(\frac{r}{r_1}\right)^{\frac{1}{2}} \chi,$$

that is, substituting for χ its value found above,

$$\psi = \frac{2q n_0}{\kappa r_1}\left\{ e^{-\iota \kappa r_1} \sin \kappa r \frac{\sin \theta_1}{1 + \mu_1} + \frac{e^{-\iota \kappa r_1}}{\sin \theta_1}(\mu_1 \sin \kappa r + \iota \cos \kappa r) - \frac{\iota e^{-\iota \kappa R}}{\sin \theta_1}\right\},$$

whence

$$\psi = \frac{2q n_0 \iota}{\kappa r_1 \sin \theta_1}\left\{ e^{-\iota \kappa (r+r_1)} - e^{-\iota \kappa R}\right\}.$$

The same value of ψ is found when $r > r_1$, and therefore all along the wire

$$\psi = \frac{2qn_0\iota}{\kappa r_1 \sin\theta_1}\left\{e^{-\iota\kappa(r+r_1)} - e^{-\iota\kappa\sqrt{(r^2+r_1{}^2+2rr_1\cos\theta_1)}}\right\}\ldots(10).$$

This result can be simply interpreted; the term

$$-\frac{2qn_0\iota}{\kappa r_1 \sin\theta_1}e^{-\iota\kappa\sqrt{(r^2+r_1{}^2+2rr_1\cos\theta_1)}}$$

represents the ψ at the point of the wire at a distance r from its end due to the source, and the term

$$\frac{2qn_0\iota}{\kappa r_1 \sin\theta_1}e^{-\iota\kappa(r+r_1)}$$

represents the effect of the free end of the wire. It ought to be observed that the electric force perpendicular to the wire, being given by

$$\frac{V^2}{\iota\kappa}\cdot\frac{1}{r\sin\theta}\frac{\partial\psi}{\partial r},$$

is large compared with ψ and, since $x\log x$ tends to zero with x, large compared with the electric force in the incident waves.

65. Taking now the case where the sources are close to the wire, that is where θ_1 is nearly equal to π, the first term of the expressions for ψ in equations (8) and (8′) is of the same order as the others and is given by

$$\frac{2qn_0}{\kappa r_1 \sin\theta_1}e^{-\iota\kappa r_1}\sin\kappa r(1-\mu),$$

when $r < r_1$. The remaining part of the series is

$$\frac{2qn_0}{\sin\theta_1}\sum_1^\infty\left(\frac{r}{r_1}\right)^{\frac{1}{2}}J_{k+\frac{1}{2}}(\kappa r)\left\{J_{-k-\frac{1}{2}}(\kappa r_1) - e^{(k+\frac{1}{2})\pi\iota}J_{k+\frac{1}{2}}(\kappa r_1)\right\}$$
$$\frac{2k+1}{2k(k+1)}\cdot(1-\mu^2)\frac{\partial P_k}{\partial\mu},$$

when $r < r_1$, which as in § 64 is equivalent to

$$\frac{2qn_0}{\kappa r_1\sin\theta_1}\left\{e^{-\iota\kappa r_1}(\mu\sin\kappa r + \iota\cos\kappa r) - \iota e^{-\iota\kappa R}\right\},$$

where $R^2 = r^2 + r_1{}^2 + 2rr_1\cos\theta;$

and therefore, in this case, ψ, for all values of μ, is given by

$$\psi = \frac{2qn_0\iota}{\kappa r_1 \sin \theta_1} \{e^{-\iota\kappa(r+r_1)} - e^{-\iota\kappa R}\},$$

when $r < r_1$, and likewise when $r > r_1$, that is

$$\psi = \frac{2qn_0\iota}{\kappa r_1 \sin \theta_1} \{e^{-\iota\kappa(r+r_1)} - e^{-\iota\kappa\sqrt{(r^2+r_1{}^2+2rr_1\cos\theta)}}\}.$$

The first term of this expression represents the effect of the free end and the second the effect of the source, the two parts being now of the same order at all points.

66.   The preceding analysis can be easily modified so as to apply to the case where there is a sphere at the end of the wire, the wire passing through the centre of the sphere. The radius of the sphere being r_0, the condition to be satisfied by ψ at the surface of the sphere is

$$\frac{\partial\psi}{\partial r} = 0,$$

when $r = r_0$.   Hence the equation, which corresponds to (8), is now

$$\psi = -q\Sigma\left(\frac{r}{r_1}\right)^{\frac{1}{2}}\left[J_{n+\frac{1}{2}}(\kappa r) - K_{n+\frac{1}{2}}(\iota\kappa r)\frac{\dfrac{\partial}{\partial r_0}\{r_0{}^{\frac{1}{2}}J_{n+\frac{1}{2}}(\kappa r_0)\}}{\dfrac{\partial}{\partial r_0}\{r_0{}^{\frac{1}{2}}K_{n+\frac{1}{2}}(\iota\kappa r_0)\}}\right]$$

$$\cdot\{J_{-n-\frac{1}{2}}(\kappa r_1) - e^{(n+\frac{1}{2})\pi\iota}J_{n+\frac{1}{2}}(\kappa r_1)\}\frac{\pi}{N\sin(n+\frac{1}{2})\pi}\sin\theta_1\frac{\partial P_n}{\partial\mu_1}(1-\mu^2)\frac{\partial P_n}{\partial\mu},$$

when $r < r_1$, with a similar expression when $r > r_1$.   When r_0 is small compared with the wave length

$$\frac{\dfrac{\partial}{\partial r_0}\{r_0{}^{\frac{1}{2}}J_{n+\frac{1}{2}}(\kappa r_0)\}}{\dfrac{\partial}{\partial r_0}\{r_0{}^{\frac{1}{2}}K_{n+\frac{1}{2}}(\iota\kappa r_0)\}}$$

is of the order $(\kappa r_0)^{2n+1}$ when n has any of its values which is not n_0, and of the order $\dfrac{1}{n_0}(\kappa r_0)^{2n+1}$ when n has the value n_0, so that, when r_0 is sufficiently small, the terms due to the presence of the sphere are negligible in comparison with the others, and therefore the effect of a small sphere at the end of the wire is negligible. The terms, which are due to the presence of the sphere, represent reflexion of the waves by the sphere, and, in practice, the conditions to be satisfied, in order that this should be negligible, will be that the radius of the sphere should be small compared with the wave length and not too great compared with the radius of the cross section of the wire.

67. The results of §§ 64, 65 can be obtained by a much simpler process, although the more direct analysis given above is preferable in the general case and necessary in the case of the investigation of § 66. It has been shewn that the waves associated with a closed circuit are permanent, and that therefore, in the case of an open circuit, radiation takes place from the free end only. The value, at any point on the straight wire, of ψ is therefore made up of two parts, one due to the source and the other to the free end, and these two together must be such that ψ vanishes at the free end. Taking the wire as axis of z, the value of ψ at any point, if there were no wire, is given, § 60, by

$$\psi \sin \phi = \frac{q\rho}{2\pi} \int_0^{2\pi} \frac{e^{-\iota\kappa R}}{R} \sin \phi_1 d\phi_1,$$

where ρ is the distance of the point from the wire, the circle of sources is chosen to be in the plane $z = 0$, and

$$R^2 = z^2 + \rho^2 + \rho_1{}^2 - 2\rho\rho_1 \cos(\phi - \phi_1),$$

where ρ_1 is the radius of the circle of sources. This is equivalent to

$$\psi = \frac{q\rho}{2\pi} \int_0^{2\pi} \frac{e^{-\iota\kappa R}}{R} \cos \chi d\chi,$$

where

$$R^2 = z^2 + \rho^2 + \rho_1{}^2 - 2\rho\rho_1 \cos \chi.$$

When there is a wire, the value of ψ along it is found by taking ρ in the above expression to be a small quantity. Writing

$$R_1{}^2 = z^2 + \rho_1{}^2,$$

it follows that

$$\frac{e^{-\iota\kappa R}}{R} = \frac{e^{-\iota\kappa R_1 + \iota\kappa \frac{\rho\rho_1}{R_1}\cos\chi}}{R_1\left(1 - \frac{\rho\rho_1}{R_1^2}\cos\chi\right)},$$

and

$$\int_0^{2\pi} \frac{e^{-\iota\kappa R}}{R}\cos\chi\, d\chi = \frac{\rho\rho_1 e^{-\iota\kappa R_1}}{R_1^2}\left(\iota\kappa + \frac{1}{R_1}\right)\int_0^{2\pi}\cos^2\chi\, d\chi,$$

that is

$$\int_0^{2\pi} \frac{e^{-\iota\kappa R}}{R}\cos\chi\, d\chi = \frac{\pi\rho\rho_1}{R_1^2} e^{-\iota\kappa R_1}\left(\iota\kappa + \frac{1}{R_1}\right),$$

and therefore the value of ψ, in the absence of the wire, is

$$\frac{q\rho^2\rho_1}{2R_1^3} e^{-\iota\kappa R_1}\left(\iota\kappa + \frac{1}{R_1}\right).$$

Hence, if Z_1 denote the electric force along the wire due to the sources,

$$\frac{\dot{Z}_1}{V^2} = \frac{q\rho_1}{R_1^2} e^{-\iota\kappa R_1}\left(\iota\kappa + \frac{1}{R_1}\right);$$

now, § 44,

$$\dot{Z}_1 = -2\log\rho_0\left(\frac{\partial^2 L}{\partial z^2} + \kappa^2 L\right),$$

where ρ_0 is the radius of the wire and, § 13,

$$L = -\frac{\iota V}{2\kappa}\psi',$$

ψ' being the value of ψ along the wire due to the source and the wire; therefore

$$\frac{\partial^2\psi'}{\partial z^2} + \kappa^2\psi' = -\frac{q\rho_1}{\log\rho_0}\frac{e^{-\iota\kappa R_1}}{R_1^2}\left(\iota\kappa + \frac{1}{R_1}\right),$$

the complete integral of which is given by

$$\psi' = A e^{\iota\kappa z} + B e^{-\iota\kappa z} - \frac{\iota q}{\kappa\rho_1\log\rho_0} e^{-\iota\kappa R_1}.$$

When the wire is indefinitely extended and there are no waves along it other than those due to the sources, the value of ψ' is given by

$$\psi' = -\frac{\iota q\, e^{-\iota\kappa R_1}}{\kappa\rho_1\log\rho_0}.$$

When the wire is terminated in one direction, taking the end as origin and measuring along the wire, the value of ψ', at a point at a distance r from the end, will be

$$\psi' = A e^{-\iota\kappa r} - \frac{\iota q\, e^{-\iota\kappa R_1}}{\kappa\rho_1 \log \rho_0},$$

where, as above, R_1 is the distance of the point on the wire from any point on the circle of sources. The condition, that ψ' should vanish at the free end, requires that

$$A = \frac{\iota q\, e^{-\iota\kappa r_1}}{\kappa\rho_1 \log \rho_0},$$

where r_1 is the distance of the free end from any point on the circle of sources, and therefore

$$\psi' = \frac{\iota q}{\kappa\rho_1 \log \rho_0} \left\{ e^{-\iota\kappa (r+r_1)} - e^{-\iota\kappa R_1} \right\},$$

which is the same result as that obtained § 64. The value of ψ' being known along the wire, the electric force at any point due to the wire can be obtained by the results of § 44, and the total electric force at any point would then be obtained by adding to this the electric force due to the sources. The effect of any disturbance, symmetrical with respect to the wire, can be immediately deduced from the above. Let q, which expresses the disturbance on the circle defined by r_1, ρ_1, be $f(t)$, some function of the time, which is known when the disturbance is given, then the value of ψ' along the wire due to this disturbance at time t is

$$\frac{1}{V\rho_1 \log \rho_0} \int^t \left\{ f\left(t - \frac{R_1}{V}\right) - f\left(t - \frac{r+r_1}{V}\right) \right\} dt.$$

CHAPTER X.

STATIONARY WAVES IN OPEN CIRCUITS.

68. In the preceding chapter the waves produced in an open circuit by a given distribution of sources were considered, the case which arises in practice is that in which stationary waves are being maintained in such a circuit or, having been set up in it, are radiating away from free ends; the circumstances of this case will be considered in this chapter.

When the nature of the radiation from a free end is known, the value of ψ along the circuit can be obtained from the results of the previous chapter, and hence the position of the nodes and the loops all along the circuit. It has been shewn, § 64, that the value of ψ at a point on a straight wire, due to a circle of sources at $r = r_1$, $\theta = \theta_1$, is given by

$$\psi = \frac{2qn_0\iota}{\kappa r_1 \sin \theta_1} \{e^{-\iota\kappa(r+r_1)} - e^{-\iota\kappa\sqrt{(r^2+r_1{}^2+2rr_1\cos\theta_1)}}\}.$$

When the circle of sources is at an indefinitely great distance from the free end of the wire, that is, when the incident waves are symmetrical with respect to the wire, and make an angle $\pi - \theta_1$ with it, the above expression takes the form

$$\psi = \frac{A\iota}{\sin\theta_1} (e^{-\iota\kappa r} - e^{-\iota\kappa r\cos\theta_1}).$$

If, through the free end of the wire, a plane be drawn perpendicular to the wire, and this plane be supposed to be perfectly conducting, the value of ψ along the wire is now obtained by adding, to the above expression, the corresponding one due to the image of the circle of sources in the plane, and is therefore given by

$$\psi = \frac{A\iota}{\sin \theta_1} \{\cos \kappa r - \cos (\kappa r \cos \theta_1)\}.$$

This result also follows from the expressions (8) and (8'), § 62, it being remembered that, when μ is approximately zero, $P_n(\mu)$ vanishes for the values of n which are given by

$$n = n_0 + 2k + 1,$$

where k is an integer, and, the wire being very thin, n_0 is very small.

If, on the system of waves diverging from the circle of sources, there be superposed a system of waves, of the same period and amplitude, converging to the circle, there results a system of stationary waves, and in this case the value of ψ along the wire is given by

$$\psi = \frac{A}{\sin \theta_1} \{\sin \kappa r - \sin (\kappa r \cos \theta_1)\}.$$

From this it follows that, in the case of stationary waves along the straight wire, the part of ψ, which corresponds to radiation from the free end in the direction making an angle $\pi - \theta_1$ with the wire, is given by

$$\psi = \frac{A}{\sin \theta_1} \{\sin \kappa r - \sin (\kappa r \cos \theta_1)\},$$

the system of sources, necessary to maintain the waves along the wire stationary, being taken into account. It appears from § 57 that, at the same distance from a free end, measured along the circuit, L, in the case of stationary waves in the circuit, is the same for any circuit, whose curvature is everywhere continuous, as for a straight circuit, if the circumstances of the radiation from the free end are the same in the two cases. Hence, in the case of any open circuit, the part of L, at any point on the circuit, which corresponds to radiation from the free end in a direction making an angle $\pi - \theta_1$ with the tangent to the circuit at the free end, this tangent being drawn in the direction of the circuit, is given by

$$L = \frac{B}{\sin \theta_1} \{\sin \kappa s - \sin (\kappa s \cos \theta_1)\},$$

where s is the distance of the point from the free end measured along the circuit, and B includes the time factor. Considering the surface, which is formed by supposing a small sphere to move with its centre on the circuit and to come to rest when the centre has reached the free end, there is no radiation across the tubular part of the surface, all the radiation takes place across the surface of that hemisphere, which closes the tube and does not intersect the circuit. Therefore the radiation from the free end of the circuit takes place in the directions which make, with the tangent to the circuit at the free end, this tangent being drawn in the direction of the circuit, angles which lie between $\pi/2$ and π. The value of L at any point of the circuit is therefore given by

$$L = \int_0^{\frac{\pi}{2}} \{\sin \kappa s - \sin (\kappa s \cos \theta_1)\} f(\theta_1) \, d\theta_1,$$

where

$$B = f(\theta_1) \sin \theta_1.$$

When there are no sources or any other free end of a circuit near to the free end, the radiation will be the same in all directions, and therefore the distribution of magnetic discontinuities will be uniform over the surface of the infinitely distant hemisphere. Now B is proportional to q, and § 60, q is proportional to $\omega \sin \theta_1 \, d\theta_1$; hence, in this case, since ω is constant, $f(\theta_1)$ is constant, and therefore L is given by

$$L = C \int_0^{\frac{\pi}{2}} \{\sin \kappa s - \sin (\kappa s \cos \theta_1)\} \, d\theta_1.$$

When there is another free end of a circuit near to the free end, the waves from the two free ends will, if of the same period, interfere, and the resultant radiation from either free end will depend on the difference of phase of the radiation from the two free ends and on the direction of the circuit or circuits at the two free ends.

69. In the case of a resonator the directions of the circuit at its two free ends are opposite, and, for the waves observed by means of the sparks which pass between the two free ends, the radiations are in opposite phases; hence the circumstances of the radiation are the same as those of the radiation from a Hertzian oscillator. To obtain the value of $f(\theta_1)$ for this case, it is necessary to find the distribution of magnetic discontinuity which is equivalent to the oscillator, this distribution being over the surface of a sphere, whose centre is at the oscillator. The value of ψ due to a circle of sources at $r = r_1$, $\theta = \theta_1$, is, § 60, given by

$$\psi = \frac{q r^{\frac{1}{2}} e^{\frac{\pi \iota}{4}}}{r_1^{\frac{1}{2}}} \sum_1^\infty e^{\frac{n\pi\iota}{2}} \frac{2n+1}{n(n+1)} J_{n+\frac{1}{2}}(\kappa r_1) K_{n+\frac{1}{2}}(\iota\kappa r) \sin^2\theta \sin\theta_1 \frac{\partial P_n}{\partial \mu} \frac{\partial P_n}{\partial \mu_1},$$

where $r > r_1$, and for a distribution of magnetic discontinuity on the surface of the sphere $r = r_1$ given by $q = g(\theta_1)\,d\mu_1$, by

$$\psi = \frac{r^{\frac{1}{2}} e^{\frac{\pi\iota}{4}}}{r_1^{\frac{1}{2}}} \sum_1^\infty e^{\frac{n\pi\iota}{2}} \frac{2n+1}{n(n+1)} J_{n+\frac{1}{2}}(\kappa r_1) K_{n+\frac{1}{2}}(\iota\kappa r) \sin^2\theta$$
$$\cdot \frac{\partial P_n}{\partial \mu} \int_{-1}^1 \sin\theta_1 \frac{\partial P_n}{\partial \mu_1} g(\theta_1)\,d\mu_1.$$

For a Hertzian oscillator ψ is given by

$$\psi = A e^{\frac{3\pi\iota}{4}} r^{\frac{1}{2}} K_{\frac{3}{2}}(\iota\kappa r) \sin^2\theta;$$

hence in the above

$$g(\theta_1) = C' \sin\theta_1,$$

giving

$$\psi = C' \frac{r^{\frac{1}{2}} e^{\frac{3\pi\iota}{4}}}{r_1^{\frac{1}{2}}} J_{\frac{3}{2}}(\kappa r_1) K_{\frac{3}{2}}(\iota\kappa r) \sin^2\theta,$$

and therefore the equivalent distribution of magnetic discontinuity over the surface of the sphere is given by

$$q = -C' \sin^2\theta_1\,d\theta_1.$$

Hence, for the waves in a resonator, which are observed by means of the sparks,

$$B = C \sin\theta_1\,d\theta_1,$$

where C only involves the time, and therefore

$$L = C \int_0^{\frac{\pi}{2}} \{\sin\kappa s - \sin(\kappa s \cos\theta_1)\} \sin\theta_1\,d\theta_1,$$

that is $$L = C\left\{\sin \kappa s - \frac{1 - \cos \kappa s}{\kappa s}\right\}.$$

70. The electric force perpendicular to the circuit, at any point on it, is proportional to $\frac{\partial L}{\partial s}$, that is, to

$$\kappa C\left\{\cos \kappa s - \frac{\sin \kappa s}{\kappa s} + \frac{1 - \cos \kappa s}{\kappa^2 s^2}\right\}.$$

The points at which there are nodes are the points at which the electric force vanishes, that is, the points at which

$$\cos \kappa s - \frac{\sin \kappa s}{\kappa s} + \frac{1 - \cos \kappa s}{\kappa^2 s^2} = 0 \dots\dots\dots\dots(1).$$

Writing $\kappa s = x$, the equation whose roots are required is

$$\cos x - \frac{\sin x}{x} + \frac{1 - \cos x}{x^2} = 0\dots\dots\dots\dots(1');$$

and this equation is equivalent to the pair

$$\cos x = \frac{\sqrt{(x^2 - 1)}}{x^2 + \sqrt{(x^2 - 1)}} \dots\dots\dots\dots (2),$$

$$\cos x = \frac{-\sqrt{(x^2 - 1)}}{x^2 - \sqrt{(x^2 - 1)}} \dots\dots\dots\dots (3).$$

The roots of (2) and (3) occur alternately, the least root of (2) being less than the least root of (3). The nth root of (2) in ascending order of magnitude can be shewn to be given by

$$x_n = 2(n - 1)\pi + \pi/2 - \xi \dots\dots\dots\dots (4),$$
where

$$\xi = \sin^{-1}\{f_n(0)\} + \frac{1}{2!}\frac{\partial}{\partial \xi}\{\sin^{-1}f_n(\xi)\}^2 + \dots,$$

$f_n(\xi)$ denoting the value of $\sqrt{(x^2 - 1)}/\{x^2 + \sqrt{(x^2 - 1)}\}$, when for x is written $2(n - 1)\pi + \pi/2 - \xi$. The nth root of (3) in ascending order of magnitude is given by

$$x_n' = (4n - 1)\pi/2 - \xi' \dots\dots\dots\dots (5),$$
where

$$\xi' = \sin^{-1}\{f_n'(0)\} + \frac{1}{2!}\frac{\partial}{\partial \xi}\{\sin^{-1}f_n'(\xi')\}^2 + \dots,$$

$f_n'(\xi')$ denoting the value of $-\sqrt{(x^2 - 1)}/\{x^2 - \sqrt{(x^2 - 1)}\}$, where for x is written $(4n - 1)\pi/2 - \xi'$. The roots of equations (2) and (3) can, with the exception of the first root in each case, be quickly calculated from the formulae (4) and (5), the first two terms of the series for ξ and ξ' giving a sufficient approximation. To obtain the values of x_1 and x_1', it is more convenient to use the equation (1'), the formulae (4) and (5) being used in these cases to indicate the neighbourhood of the roots required. The formulae (4) and (5) shew that the least root of equation (2) is in the neighbourhood of $71\pi/180$, and that the least root of equation (3) is in the neighbourhood of $253\pi/180$. Using the tables of the circular functions, and writing

$$\cos x - \frac{\sin x}{x} + \frac{1 - \cos x}{x^2} = X,$$

it appears that, when $x = 71\frac{2}{15}\pi/180$, $X = \cdot0001593...$, and when $x = 71\frac{3}{20}\pi/180$, $X = -\cdot0000305...$; hence the least root of equation (1') lies between $1067\pi/2700$ and $1423\pi/3600$. Again, when $x = 253\frac{3}{5}\pi/180$, $X = -\cdot0001473...$, and when $x = 253\frac{3}{60}\pi/180$, $X = \cdot0001127...$; hence the next root of equation (1') lies between $1268\pi/900$ and $15217\pi/10800$. The third root of equation (1') is approximately $1663\pi/675$, and so on.

71. For any resonator, the fundamental wave length is that for which there is only one node on the resonator, that node being at the point which is equidistant from the two ends, the distances being measured along the circuit. Hence if l is the length of the resonator, the fundamental wave length λ_0 satisfies the relation

$$\frac{1423\pi}{3600} > \frac{2\pi}{\lambda_0} \cdot \frac{l}{2} > \frac{1067\pi}{2700},$$

that is,

$$\frac{1423}{3600} > \frac{l}{\lambda_0} > \frac{1067}{2700},$$

or

$$2\cdot5328 > \frac{\lambda_0}{l} > 2\cdot52928.$$

Therefore $\lambda_0 = 2\cdot53l$, the difference between this expression for λ_0 and the accurate one being less than $\cdot04$ per cent. The wave length, corresponding to the first overtone, is that for

which there are three nodes on the resonator; denoting this
wave length by λ_1, it follows from the above that

$$\frac{15217}{10800} > \frac{l}{\lambda_1} > \frac{1268}{900},$$

that is, $\cdot709779 > \frac{\lambda_1}{l} > \cdot709732$;

and therefore $\lambda_1 = \cdot7097l$. Similarly the wave length λ_2,
corresponding to the second overtone, is given by $\lambda_2 = \cdot4059l$,
the values of the wave lengths corresponding to the successive
overtones tending to the values given by $\lambda_n = 2l/(2n + 1)$, as n
increases. The relations between the successive wave lengths
are $\lambda_1 = \cdot280\lambda_0$, $\lambda_2 = \cdot160\lambda_0 = \cdot572\lambda_1$, etc.

For a circular resonator, neglecting the gap, $l = \pi D$, where
D is the diameter of the resonator, and the fundamental wave
length in this case is, by the above, given by $\lambda_0 = 7\cdot95D$, this
expression differing from the accurate one by less than $\cdot025$
per cent.

72. It has been shewn experimentally by Sarasin and
de la Rive* that, when electric waves are being propagated
along a wire with a free end, the waves detected by a resonator
are those which have the same period as the oscillations which
belong to the resonator. The resonators used by them were
circular, their diameters being $\cdot75$, $\cdot50$, and $\cdot35$, in metres
respectively, and the observed distances between the first and
second nodes along the wire from the free end were $2\cdot95$, $1\cdot95$
and $1\cdot43$. It will be shewn later, §76, that the distance between
the first and second nodes is very approximately half a wave
length, so that the observed distances are approximately half
the fundamental wave lengths of the corresponding resonators.
The values of the half wave lengths of the resonators, as
calculated from the formula $\lambda_0 = 7\cdot95D$, are $2\cdot98$, $1\cdot98$ and $1\cdot38$
respectively, which agree well with the observed values.

In their later experiments† the wave length in air was
observed by detecting the nodes and loops in the stationary

---

* *Comptes Rendus*, cx. 1890.
† *Comptes Rendus*, cxii. 1891.

waves due to reflexion at a large plane reflector. In discussing these experiments the distance of the first node from the reflector will be taken to be half the wave length, it being easier to observe accurately a node than a loop, and the position of the first node being less affected than that of the second by the finite extent of the reflector.

Diameter of reso- nator circle	1 metre stout wire 1 cm. in diameter	·75 m. stout wire	·50 m. stout wire	·35 m. stout wire	·35 m. fine wire 2 mm. diameter	·25 m. stout wire	·25 m. fine wire	·20 m. stout wire	·20 m. fine wire	·10 m. stout wire
Dist. of first node	4·14	3·01	2·22*	1·49	1·51	·94	1·17	·80	·93	·41
$\frac{\lambda}{4}$ observed value	2·07	1·50	1·11	·74	·75	·47	·58	·40	·46	·20
$\frac{\lambda}{4}$ calculated value	1·98	1·49	·99	·69	·69	·49	·49	·39	·39	·19

* No first node observed, only the first loop.

In the above table the calculated values of the quarter wave lengths have been obtained from the formula $\lambda = 7·95D$, and it will be seen that the agreement between calculated and observed values is in general very close. These experiments were repeated by Sarasin and de la Rive† under more favourable conditions. The results obtained for circular resonators of diameters ·75 m. and ·50 m. were that the distance between a loop and a node in the two cases were 1·50 m. and 1·00 m. respectively. These observed values give the formula $\lambda = 8D$ for the wave length, the difference between which and the formula which results from the theory being about ·6 per cent.

73. In the preceding investigation of the wave lengths of resonators, it has been assumed that the effect of the small spheres, which are placed at the ends to make the sparks more definite, is negligible, and also that the damping of the oscillations in the resonator is negligible. The expression for ψ in the case of a straight wire, with a sphere of radius r_0 at the free end, under the influence of a circle of magnetic discontinuities,

† *Comptes Rendus*, cxv. 1892.

has been given, § 66, and it appears from it that the difference, between the term involving n in this case and the term involving n in the case when there is no sphere, is of the order $(r_0/\lambda)^{2n+1}$, when n has any of its values other than n_0. The most effective of these terms is that for which $n = n_0 + 1$, and in this case the ratio of the part, due to a sphere of radius ·01 m., to the whole term lies between 10^{-5} and 10^{-6} for oscillations whose wave length is 2 m. The spheres at the ends of resonators have diameters lying between 1 cm. and 2 cm., so that, for a circular resonator of ·25 m. diameter, the effect of the small spheres, as far as any term in the series for ψ, and therefore for L, other than the first, is concerned, is negligible, and the same result will hold for larger resonators. The value of ψ for the straight wire, with a small sphere of radius r_0 at the free end, under the influence of a circle of magnetic discontinuity at $r = r_1$, $\theta = \theta_1$, is therefore, § 66, given by

$$\psi = - q\,\frac{r^{\frac{1}{2}}}{r_1^{\frac{1}{2}}} \left[J_{n_0+\frac{1}{2}}(\kappa r) - K_{n_0+\frac{1}{2}}(\iota\kappa r)\,\frac{\dfrac{\partial}{\partial r_0}\{r_0^{\frac{1}{2}}J_{n_0+\frac{1}{2}}(\kappa r_0)\}}{\dfrac{\partial}{\partial r_0}\{r_0^{\frac{1}{2}}K_{n_0+\frac{1}{2}}(\iota\kappa r_0)\}} \right]$$

$$.\,\big[J_{-n_0-\frac{1}{2}}(\kappa r_1) - e^{(n_0+\frac{1}{2})\pi\iota}J_{n_0+\frac{1}{2}}(\kappa r_0)\big]\,\frac{\pi}{N_0\sin(n_0+\frac{1}{2})\pi}$$

$$.\,\sin\theta_1(1-\mu^2)\frac{\partial P_{n_0}}{\partial\mu_1}\frac{\partial P_{n_0}}{\partial\mu}$$

$$- q\,\frac{r^{\frac{1}{2}}}{r_1^{\frac{1}{2}}}\sum_{n=n_0+1}^{\infty} J_{n+\frac{1}{2}}(\kappa r)\{J_{-n-\frac{1}{2}}(\kappa r_1) - e^{(n+\frac{1}{2})\pi\iota}J_{n+\frac{1}{2}}(\kappa r_1)\}\frac{\pi}{N\sin(n+\frac{1}{2})\pi}$$

$$.\,\sin\theta_1(1-\mu^2)\frac{\partial P_n}{\partial\mu_1}\frac{\partial P_n}{\partial\mu}.$$

As in § 64, the value of ψ along the wire is given by

$$\psi = \frac{2qn_0\iota}{\kappa r_1\sin\theta_1}\{e^{-\iota\kappa(r+r_1)} - e^{-\iota\kappa\sqrt{(r^2+r_1^2+2rr_1\cos\theta_1)}}\}$$

$$+ \frac{qr^{\frac{1}{2}}}{r_1^{\frac{1}{2}}}K_{n_0+\frac{1}{2}}(\iota\kappa r)\,\frac{\dfrac{\partial}{\partial r_0}\{r_0^{\frac{1}{2}}J_{n_0+\frac{1}{2}}(\kappa r_0)\}}{\dfrac{\partial}{\partial r_0}\{r_0^{\frac{1}{2}}K_{n_0+\frac{1}{2}}(\iota\kappa r_0)\}}\{J_{-n_0-\frac{1}{2}}(\kappa r_1) - e^{(n_0+\frac{1}{2})\pi\iota}J_{n_0+\frac{1}{2}}(\kappa r_1)\}$$

$$.\,\frac{\pi}{N_0\sin(n_0+\frac{1}{2})\pi}\sin\theta_1\frac{\partial P_{n_0}}{\partial\mu_1}(1-\mu^2)\frac{\partial P_{n_0}}{\partial\mu},$$

that is, by

$$\psi = \frac{2qn_0\iota}{\kappa r_1 \sin \theta_1} \left[e^{-\iota\kappa (r+r_1)} \left\{1 - \iota\nu (1 - \mu_1)\right\} - e^{-\iota\kappa \sqrt{(r^2+r_1^2+2rr_1 \cos \theta_1)}} \right],$$

where

$$\nu = \frac{2n_0}{N_0 \sin (n_0 + \tfrac{1}{2}) \pi} \frac{\dfrac{\partial}{\partial r_0} \left\{r_0^{\frac{1}{2}} J_{n_0+\frac{1}{2}} (\kappa r_0)\right\}}{\dfrac{\partial}{\partial r_0} \left\{r_0^{\frac{1}{2}} J_{-n_0-\frac{1}{2}} (\kappa r_0) - e^{(n_0+\frac{1}{2})\pi\iota} r_0^{\frac{1}{2}} J_{n_0+\frac{1}{2}} (\kappa r_0)\right\}}.$$

When κr_0 is small compared with n_0, ν tends to zero, and when n_0 is small compared with κr_0, ν tends to the value $-\iota e^{\iota\kappa r_0} \cos \kappa r_0$. The value of ν in general is $\epsilon - \iota\epsilon'$, where the absolute value of ν is always less than unity. When κr_0 is so small compared with n_0, that $(\kappa r_0/n_0)^2$ can be neglected, $\nu = \epsilon$, and the value of L for a resonator under these conditions is given by

$$L = A \int_0^1 \left\{\sin \kappa s + \epsilon (1 - \mu_1) \cos \kappa s - \sin \kappa s\mu_1\right\} d\mu_1,$$

that is, by

$$L = A \left\{\sin \kappa s + \frac{\epsilon}{2} \cos \kappa s - \frac{1 - \cos \kappa s}{\kappa s}\right\}.$$

The nodes are therefore determined by the roots of the equation

$$\cos \kappa s - \frac{\epsilon}{2} \sin \kappa s - \frac{\sin \kappa s}{\kappa s} + \frac{1 - \cos \kappa s}{\kappa^2 s^2} = 0\ldots\ldots\ldots(6).$$

Denoting by x_0 the least root of equation (1′), which is what (6) becomes, when $\epsilon = 0$, the least root of equation (6) is approximately $x_0 - \cdot 7\epsilon$. Hence, if ϵ is less than 10^{-3}, the fundamental wave length of the resonator is increased by less than $\cdot 1$ per cent. In general, when $\nu = \epsilon - \iota\epsilon'$, the value of L for a resonator is given by

$$L = A \int_0^1 \left[\left\{1 - \epsilon' (1 - \mu_1)\right\} \sin \kappa s + \epsilon (1 - \mu_1) \cos \kappa s - \sin \kappa s\mu_1\right] d\mu_1,$$

that is, by

$$L = A \left\{\left(1 - \frac{\epsilon'}{2}\right) \sin \kappa s + \frac{\epsilon}{2} \cos \kappa s - \frac{1 - \cos \kappa s}{\kappa s}\right\}.$$

The nodes are therefore determined by the roots of the equation

$$\left(1 - \frac{\epsilon'}{2}\right)\cos \kappa s - \frac{\epsilon}{2}\sin \kappa s - \frac{\sin \kappa s}{\kappa s} + \frac{1 - \cos \kappa s}{\kappa^2 s^2} = 0 \dots (7).$$

The least root of this equation is less than the least root of equation (1), and therefore the fundamental wave length of the resonator is greater, that is, the effect of the small spheres at the ends of the resonator is always to increase the fundamental wave length. In their second series of experiments* Sarasin and de la Rive used resonators, which were made of two different kinds of wire, one being stout wire of diameter 1 cm., the other fine wire of diameter 2 mm. The radius of the small spheres being about 1 cm., the reflexion of the waves due to the small spheres, in the case of the stout wire, will be small, but, in the case of the fine wire, appreciable. It might therefore be expected, that the wave lengths of a resonator made from the fine wire, would be greater than that of a resonator of the same size, made from the stout wire, more especially in the case of the resonators of less diameter, as the effect also depends on the ratio r_0/λ. This difference is well marked in the results of the experiments, the increase, in the case of the resonator of ·20 m. diameter, being rather more than 15 per cent. In practice, this cause of difference could be got rid of by diminishing the amount of reflexion due to the small bodies at the ends of the wire, and this could be effected by using, instead of spheres, pear-shaped bodies, the wire being fitted on to the narrower ends of these bodies.

74. In some experiments the ends of the resonator have been fitted with small plates, instead of with spheres, and it is of some importance to find out what their effect would be. Although this problem cannot be solved directly, the effect can be very approximately determined by means of the preceding analysis. Since the result depends on the amount of the reflexion of the waves by the plate, it is clear that the effect of the plate is the same as that of a small sphere, when the

* *Comptes Rendus*, cxii. 1891.

wire is made indefinitely thin, and this is the case where, in the
above analysis, $\nu = -\iota e^{\iota \kappa r_0} \cos \kappa r_0$, κr_0 being very small. The
function L, for the resonator, is therefore given by

$$L = A \left\{ (1 - \tfrac{1}{2} \cos^2 \kappa r_0) \sin \kappa s + \tfrac{1}{2} \sin \kappa r_0 \cos \kappa r_0 \cos \kappa s - \frac{1 - \cos \kappa s}{\kappa s} \right\},$$

and the nodes are determined by the roots of the equation

$$(1 - \tfrac{1}{2} \cos^2 \kappa r_0) \cos \kappa s - \tfrac{1}{2} \sin \kappa r_0 \cos \kappa r_0 \sin \kappa s$$
$$- \frac{\sin \kappa s}{\kappa s} + \frac{1 - \cos \kappa s}{\kappa^2 s^2} = 0 \ldots (8).$$

If the wave length be supposed given, the least root of
equation (8) is very small, and therefore the distance of the
first node from the end is very small. In the case of a
resonator, oscillating in the fundamental mode, the distance of
the first node from the end is half the length of the resonator,
and the fundamental wave length is therefore very great.
Now, in a resonator, the oscillations, which are observed, are
those, whose wave length is least different from that of the
oscillator, as the intensity of the oscillations in the resonator
depends, approximately, on the inverse of the square root of the
difference of the squares of the reciprocals of the wave lengths
of the oscillator and the resonator. Hence, in this case, the
oscillations, corresponding to the fundamental wave length,
would not be observed. To determine the wave length corresponding to the first overtone, it will be sufficient to discuss,
instead of equation (8), the equation

$$\tfrac{1}{2} \cos \kappa s - \frac{\sin \kappa s}{\kappa s} + \frac{1 - \cos \kappa s}{\kappa^2 s^2} = 0 \ldots \ldots \ldots (8').$$

The least root of equation (8') is given by $\kappa s = 0$, which has
already been considered. Writing

$$\tfrac{1}{2} \cos x - \frac{\sin x}{x} + \frac{1 - \cos x}{x^2} = X,$$

it appears that, when $x = 234\frac{1}{12} \pi/180$, $X = - \cdot 000014$, and when
$x = 234\frac{1}{10} \pi/180$, $X = \cdot 000097$, hence the next root of equation
(8') lies between $234\frac{1}{12} \pi/180$ and $234\frac{1}{10} \pi/180$. Taking the
second of these two values, which will give a sufficiently

accurate result, the wave length λ_1, corresponding to the first overtone, is given by

$$\frac{l}{\lambda_1} = \frac{2341}{1800},$$

where l is the length of the resonator, that is, $\lambda_1 = \cdot7689\,l$. The wave length, corresponding to the first overtone, in the case where the effect of the small spheres at the end is negligible, was determined above, and is $\lambda_1 = \cdot7097l$, so that the wave length, corresponding to the first overtone, when plates, instead of spheres, are used at the ends of the resonator, is between 7 and 8 per cent. greater. The oscillators, which have been used in different experiments, appear to emit waves of wave lengths lying between 5 m. and 8 m., so that, when a circular resonator, of diameter 1 m. or less is used, the ends being fitted with plates, the oscillations observed will be those whose wave length is λ_1, where $\lambda_1 = 2\cdot41\,D$.

75. Limits for the rate of decay of the oscillations in a resonator can be found as follows. The resonator being, in respect of the radiation from it, equivalent to a Hertzian oscillator, the rate of radiation of energy from it, during any period, is, § 49, $16\pi^4E^2g^2V/3\lambda^4$, where g is the effective gap, E is the maximum charge at either end of the resonator during the period under consideration, λ is the wave length of the oscillations, and V is the velocity of radiation in the surrounding medium. Hence, if W denotes the total energy in the resonator at any time,

$$\frac{dW}{dt} = -\frac{16\pi^4E^2g^2V}{3\lambda^4}.$$

Now, at any instant of time, $W = \cdot E^2/l$, where l is a length, which lies between the greatest distance between any two points of the resonator and the distance between the effective free ends of the resonator; therefore

$$\frac{dW}{dt} = -\frac{16\pi^4g^2lV}{3\lambda^4}\,W,$$

and
$$W = e^{-kt}\,W_0,$$

where $k = 16\pi^4 g^2 l V/3\lambda^4$. Hence the time, which elapses before the amplitude of the oscillations falls to $1/e$ of its initial value, is $3\lambda^4/8\pi^4 g^2 l V$, that is $3\lambda^3 T/8\pi^4 g^2 l$, where T is the time of a complete oscillation. In the case of a circular resonator, of diameter D, $D > l > g$, and the time t, which elapses before the amplitude of the oscillations falls to $1/e$ of its initial value, satisfies the relation

$$3\lambda^3 T/8\pi^4 g^3 > t > 3\lambda^3 T/8\pi^4 g^2 D.$$

When the ends of the resonator are spheres, the above relation is approximately $1 \cdot 92 T (D/g)^3 > t > 1 \cdot 92 T (D/g)^2$, and, in the case of the resonator, of diameter $\cdot 70$ m., used by Hertz, the distance between the centres of the small spheres being 2 cm., the number of complete oscillations, executed before the amplitude falls to $1/e$ of its initial value, lies between 2352 and 82320. When the ends of the resonator are plates, the time t satisfies the relation $\cdot 0533 T (D/g)^3 > t > \cdot 0533 T (D/g)^2$, which, in the case of a resonator of the same size as that used by Hertz, shews that, when the ends are plates, the number of complete oscillations, executed before the amplitude falls to $1/e$ of its initial value, lies between 65 and 2253. Bjerknes[*] found, by experiment, that the number of complete oscillations, executed before the amplitude falls to $1/e$ of its initial value, was between 500 and 600, in the case of a resonator, whose fundamental wave length was about 8 metres, but, in comparing the results of his experiments with those of theory, it has to be remembered that, in his arrangement, sparks were prevented from passing between the ends of the resonator. Now, sparks can be prevented from passing in two ways, by increasing the effective gap or by arranging that there shall be nodes near the ends of the resonator, in which case the potential difference in the gap will be much less. It at once follows from the above, that an increase in the length of the effective gap greatly increases the rate of decay. The second method of preventing sparks is equivalent to preventing the resonator from executing vibrations of fundamental wave length, which can be effected by using plates at the ends, instead of spheres, and in this case,

* *Annalen der Physik und Chemie*, Bd. 44, 1891.

as was seen above, the rate of decay is about 36 times faster. It can, therefore, be concluded, that the rate of decay of the oscillations in the resonator, under the conditions of Bjerknes' experiments, is considerably greater than it is, when the resonator is arranged as in the experiments of Hertz or in those of Sarasin and de la Rive, and, taking into account the influence of the arrangement of the resonator on the rate of decay, the rate observed by Bjerknes is of the order indicated by theory. It follows, also, from the above, that the rate of decay of the oscillations in the resonator, which correspond to the first and higher overtones, is very much greater than that of the oscillations of fundamental wave length. The influence of the rate of decay on the wave length of the fundamental oscillations of the resonator will be negligible, as was assumed § 71, this rate being very small.

76. Proceeding now to the case of waves along a straight wire with a free end, there being no sources or other free ends near to the free end, it was seen, § 68, that the value of L, at a point on the wire at a distance s, measured along the wire, from the free end is given by

$$L = A \int_0^{\frac{\pi}{2}} \{\sin \kappa s - \sin (\kappa s \cos \theta)\} \, d\theta.$$

The points at which the electric force perpendicular to the wire vanishes, that is the points at which there are nodes, are determined by the roots of the equation

$$\frac{\pi}{2} \cos \kappa s - \int_0^{\frac{\pi}{2}} \cos (\kappa s \cos \theta) \cos \theta \, d\theta = 0.$$

Writing

$$\frac{\pi}{2} \cos x - \int_0^{\frac{\pi}{2}} \cos (x \cos \theta) \cos \theta \, d\theta = X,$$

it appears that, when $x = 0$, $X = \pi/2 - 1$. As x increases from zero, $\frac{\pi}{2} \cos x$ diminishes and vanishes when $x = \pi/2$, the integral

$$\int_0^{\frac{\pi}{2}} \cos (x \cos \theta) \cos \theta \, d\theta$$ also diminishes and remains positive

until its first zero, which can easily be shewn to be less than $\sqrt{3}$, is attained; hence, the least value of x, for which X vanishes, lies between 0 and $\pi/2$. Writing the integral in the expression for X in the form of a series, X is given by

$$X = \frac{\pi}{2} \cos x - \left(1 - \frac{x^2}{3} + \frac{x^4}{3^2 \cdot 5} - \frac{x^6}{3^2 \cdot 5^2 \cdot 7} + \ldots \right),$$

and, using this expression, it can be shewn that, when $x = 23\pi/60$, $X = \cdot0015\ldots$, and, when $x = 69\frac{1}{3}\pi/180$, $X = -\cdot0022\ldots$; therefore the least zero of X lies between $23\pi/60$ and $208\pi/540$. For waves of wave length λ, the first node is therefore at a distance d from the end of the wire, where

$$\frac{104\lambda}{540} > d > \frac{23\lambda}{120},$$

whence $d = \cdot192\lambda$, this result differing from the accurate one by less than one-two-thousandth part of a wave length. It can also be shewn that the next zero is greater than π, and that, when $x = 248\pi/180$, $X = -\cdot0032\ldots$, and when $x = 248\frac{1}{6}\pi/180$, $X = \cdot00027\ldots$; hence, if d_1 is the distance of the second node, d_1 satisfies the relation

$$\frac{1489\lambda}{2160} > d_1 > \frac{124}{180}\lambda,$$

and therefore $d_1 = \cdot689\lambda$, this result differing from the accurate one by less than $10^{-3}/4$. The distance between the first and second nodes is given by $\cdot497\lambda$, which differs from half a wave length by $\cdot003\lambda$, justifying the assumption made § 72. From the formula $\lambda = 7\cdot95D$, for the fundamental wave length of a circular resonator of diameter D, § 71, it follows that the semi-circumference of the circle is $\cdot197\lambda$, so that the semi-circumference of the resonator is very nearly equal to the distance of the first node from the free end of a wire along which waves are travelling, the node being that which belongs to oscillations of the same period as those of fundamental wave length in the resonator. This was observed to be the case by Sarasin and de la Rive* in their experiments. The result of taking into

* *Comptes Rendus*, cx. 1890.

account the effect of the small spheres at the ends of the resonator would be to make the coincidence closer.

To determine the positions of the other nodes along the wire, the semi-convergent form of the integral $\int_0^{\frac{\pi}{2}} \cos(x \cos\theta) \cos\theta \, d\theta$, can be used, and the expression whose zeros are required, is, in this form,

$$\frac{\pi}{2}\cos x - \sqrt{\frac{\pi}{2x}} \cos\left(x - \frac{\pi}{4}\right) \left\{1 + \frac{3 \cdot 5 \cdot 1}{2!\,(8x)^2} - \frac{3 \cdot 5 \cdot 7 \cdot 9 \cdot 1 \cdot 3 \cdot 5}{4!\,(8x)^4} + \ldots\right\}$$

$$+ \sqrt{\frac{\pi}{2x}} \sin\left(x - \frac{\pi}{4}\right) \left\{\frac{3}{8x} - \frac{3 \cdot 5 \cdot 7 \cdot 1 \cdot 3}{3!\,(8x)^3} + \ldots\right\}$$

$$+ \left(\frac{1}{x^2} - \frac{3}{x^4} + \frac{3^2 \cdot 5}{x^6} - \ldots\right).$$

The zeros of this expression ultimately tend to $(2k+1)\pi/2$, where k is an integer; the distance between the second and third nodes is greater than half a wave length, and so on.

77. The form of the wave front in the neighbourhood of the free end of a wire can be obtained as follows. Taking r in equation (8), § 62, to be very small compared with the wave length, the function ψ, in the neighbourhood of the free end, is given by

$$\psi = n_0 \, Br \, (1 - \mu),$$

and

$$\frac{\partial \psi}{\partial \mu} = - n_0 \, Br \, P_{n_0}(\mu),$$

from the same equation, whence the components of the electric force R and Θ are given by

$$\frac{1}{V^2} \frac{\partial R}{\partial t} = \frac{1}{r^2} \frac{\partial \psi}{\partial \mu} = - \frac{n_0 \, B}{r} P_{n_0}(\mu),$$

$$\frac{1}{V^2} \frac{\partial \Theta}{\partial t} = \frac{1}{r \sin\theta} \frac{\partial \psi}{\partial r} = \frac{n_0 \, B}{r} \sqrt{\frac{1 - \mu}{1 + \mu}}.$$

Therefore in the immediate neighbourhood of the free end,

$$-\frac{R}{\Theta} = P_{n_0}(\mu) \sqrt{\frac{1+\mu}{1-\mu}},$$

that is,

$$-\frac{R}{\Theta} = \sqrt{\frac{1+\mu}{1-\mu}} \left\{ 1 + n_0 \log \frac{1+\mu}{2} \right\},$$

or, if θ be the angle which the radius vector makes with the wire,

$$-\frac{R}{\Theta} = \tan\frac{\theta}{2} \left(1 + 2n_0 \log \sin\frac{\theta}{2} \right).$$

Hence the equation to the wave fronts near to the free end is approximately

$$r^{\frac{1}{2}} \cos\frac{\theta}{2} = \text{const.},$$

that is, they are portions of paraboloids of revolution, which have the free end as focus and their vertices on the wire.

78. To obtain the form of the wave front in the neighbourhood of the wire, at a great distance from the free end, let the wire be taken as the axis of z, the axes of x and y being perpendicular to it, and the origin at the free end. The expressions for the components of the electric force at the point x, y, z are then, §§ 44, 57,

$$X = \frac{\partial}{\partial x} \int_0^\infty \frac{e^{-\iota\kappa r}}{r} \frac{\partial L}{\partial z_1} dz_1,$$

$$Y = \frac{\partial}{\partial y} \int_0^\infty \frac{e^{-\iota\kappa r}}{r} \frac{\partial L}{\partial z_1} dz_1,$$

$$Z = \frac{\partial}{\partial z} \int_0^\infty \frac{e^{-\iota\kappa r}}{r} \frac{\partial L}{\partial z_1} dz_1 + \kappa^2 \int_0^\infty \frac{e^{-\iota\kappa r}}{r} L dz_1,$$

where $r^2 = (z - z_1)^2 + x^2 + y^2$, and, § 68,

$$L = C \left[\sin\kappa z_1 - \frac{2}{\pi} \int_0^{\frac{\pi}{2}} \sin(\kappa z_1 \cos\theta)\, d\theta \right].$$

Writing

$$\chi = \int_0^\infty \frac{e^{-\iota \kappa r}}{r} \frac{\partial L}{\partial z_1} \, dz_1,$$

and for $\dfrac{e^{-\iota \kappa r}}{r}$ its equivalent

$$\frac{\kappa}{\sqrt{2\pi}} \int_{c-\infty\iota}^0 e^{\frac{\sigma}{2} - \frac{\kappa^2}{2\sigma} \{(z-z_1)^2 + x^2 + y^2\}} \frac{d\sigma}{\sigma^{\frac{3}{2}}},$$

it follows that

$$\chi = \frac{-\kappa^2 C}{\sqrt{2\pi}} \int_0^\infty dz_1 \int_{c-\infty\iota}^0 e^{\frac{\sigma}{2} - \frac{\kappa^2}{2\sigma} \{(z-z_1)^2 + x^2 + y^2\}}$$

$$\cdot \frac{d\sigma}{\sigma^{\frac{3}{2}}} \left[\cos \kappa z_1 - \frac{2}{\pi} \int_0^{\frac{\pi}{2}} \cos (\kappa z_1 \cos \theta) \cos \theta \, d\theta \right].$$

Now

$$\int_0^\infty e^{-\frac{\kappa^2}{2\sigma}(z-z_1)^2} \cos \kappa z_1 \, dz_1 = \int_{-z}^\infty e^{-\frac{\kappa^2 \zeta^2}{2\sigma}} \cos \kappa (z + \zeta) \, d\zeta,$$

where z is positive, that is

$$\int_0^\infty e^{-\frac{\kappa^2}{2\sigma}(z-z_1)^2} \cos \kappa z_1 \, dz_1$$

$$= \int_{-\infty}^\infty e^{-\frac{\kappa^2 \zeta^2}{2\sigma}} \cos \kappa (z + \zeta) \, d\zeta - \int_{-\infty}^{-z} e^{-\frac{\kappa^2 \zeta^2}{2\sigma}} \cos \kappa (z + \zeta) \, d\zeta,$$

whence

$$\int_0^\infty e^{-\frac{\kappa^2}{2\sigma}(z-z_1)^2} \cos \kappa z_1 \, dz_1$$

$$= \frac{\cos \kappa z}{\kappa} e^{-\frac{\sigma}{2}} \sqrt{2\pi\sigma} - \int_{-\infty}^{-z} e^{-\frac{\kappa^2 \zeta^2}{2\sigma}} \cos \kappa (z + \zeta) \, d\zeta.$$

The second term on the right-hand side, when z is great, is negligible compared with the first, and therefore

$$\int_0^\infty e^{-\frac{\kappa^2}{2\sigma}(z-z_1)^2} \cos \kappa z_1 \, dz_1$$

tends to the value $\dfrac{1}{\kappa} \cos \kappa z \, e^{-\frac{\sigma}{2}} \sqrt{2\pi\sigma}$, when z is great. Similarly

$$\int_0^\infty e^{-\frac{\kappa^2}{2\sigma}(z-z_1)^2} \cos (\kappa z_1 \cos \theta) \, dz_1$$

tends to the value $\dfrac{1}{\kappa}\cos{(\kappa z \cos\theta)}\, e^{-\frac{\sigma \cos^2\theta}{2}}\sqrt{2\pi\sigma}$, when z is great, and therefore χ tends to the value

$$-\kappa C \int_{c-\infty\iota}^{0} e^{\frac{\sigma}{2}-\frac{\kappa^2}{2\sigma}(x^2+y^2)}$$

$$\times \left\{ e^{-\frac{\sigma}{2}}\cos\kappa z - \frac{2}{\pi}\int_{0}^{\frac{\pi}{2}} e^{-\frac{\sigma}{2}\cos^2\theta}\cos{(\kappa z \cos\theta)}\cos\theta \, d\theta \right\} \frac{d\sigma}{\sigma}.$$

Further, the first significant term of the integral

$$\int_{0}^{\frac{\pi}{2}} e^{-\frac{\sigma}{2}\cos^2\theta}\cos{(\kappa z \cos\theta)}\cos\theta \, d\theta$$

involves $(\kappa z)^{-\frac{1}{2}}$. Hence ultimately, when z is very great compared to the wave length, χ is given by

$$\chi = -\kappa C \cos\kappa z \int_{c-\infty\iota}^{0} e^{-\frac{\kappa^2}{2\sigma}(x^2+y^2)}\frac{d\sigma}{\sigma}.$$

In like manner, the integral

$$\int_{0}^{\infty} \frac{e^{-\iota\kappa r}}{r} L \, dz_1 = -C \sin\kappa z \int_{c-\infty\iota}^{0} e^{-\frac{\kappa^2}{2\sigma}(x^2+y^2)}\frac{d\sigma}{\sigma},$$

when z is very great compared to the wave length. Hence, at a great distance from the plane $z = 0$, in the direction of the wire, the components of the electric force are given by

$$X = \kappa^3 C x \cos\kappa z \int_{c-\infty\iota}^{0} e^{-\frac{\kappa^2}{2\sigma}(x^2+y^2)}\frac{d\sigma}{\sigma^2},$$

$$Y = \kappa^3 C y \cos\kappa z \int_{c-\infty\iota}^{0} e^{-\frac{\kappa^2}{2\sigma}(x^2+y^2)}\frac{d\sigma}{\sigma^2},$$

$$Z = 0;$$

that is,

$$X = -\frac{2\kappa C x}{x^2 + y^2}\cos\kappa z,$$

$$Y = -\frac{2\kappa C y}{x^2 + y^2}\cos\kappa z,$$

$$Z = 0.$$

Thus, at a distance from the plane $z = 0$, great in comparison with the wave length, the wave fronts tend to become planes perpendicular to the wire. Therefore, in travelling in the direction of the wire from the free end, the wave fronts change continuously from paraboloids of revolution with their concavities towards the free end, the free end being the common focus, to planes perpendicular to the wire.

79. The form of the wave fronts, in the neighbourhood of a terminated straight wire, has been made the subject of experimental investigation by Birkeland and Sarasin*. Circular resonators, of diameters 10 cm. and 25 cm., were used by them ; the positions of the first four nodes, in the direction of the wire from the free end, at different distances from the wire, are given in a table, for the resonator of diameter 10 cm., and a figure is given, indicating the positions of the nodes at different distances from the wire and the forms of the wave fronts, for both resonators. For the resonator of diameter 10 cm., it appears that, at a distance of 2 cm. from the wire, the first node was at a distance of 16 cm. from the free end of the wire, measured parallel to the wire, and that the mean distance between consecutive nodes was 39·6 cm.

The result of the theory, § 71, is that the wave length of the resonator is 79·5 cm., and, § 76, that the distance of the first node on the wire from the free end, for the waves belonging to a resonator of 10 cm. diameter, is 15·2 cm., the distance between successive nodes being approximately half a wave length, which in this case is 39·7 cm., so that the results of theory and observation agree. In the figure given by Birkeland and Sarasin, the change in the form of the wave front, as the free end is receded from in the direction of the wire, is the same as that arrived at theoretically §§ 77, 78. In particular, they found in the case " of the three sets of observations made in the spaces between the nodes, that the perpendicular to the plane of the resonator circle, when placed in that position, in which the effect was a maximum, is very approximately in the direction of the bisector

---

* *Comptes Rendus*, cxvii. 1893.

of the angle, which the line joining it to the end of the wire makes with a parallel to the wire through it." When the spark gap is in the position described, it is tangential to the wave front, and therefore the wave front is very approximately a paraboloid of revolution, having the end of the wire as focus, which is in exact agreement with the result of the theory, § 77.

80. When the velocity of propagation of electric waves is determined by observations on wires, the thing measured is the distance between two consecutive nodes along the wire, and this distance is assumed to be half a wave length. Now the result of theory is that, if the wire were perfectly conducting and endless, the wave fronts would, for a straight wire, be planes perpendicular to the wire, and the distance between two consecutive nodes along the wire would be half a wave length, so that the velocity of propagation determined under these conditions would be the accurate one.

If the wire is not perfectly conducting, and the effect of imperfect conduction is assumed to be dissipation of the energy according to Ohm's law, the wave fronts, in the case of a straight wire, will no longer be planes perpendicular to the wire, so that the distance between two consecutive nodes along the wire will not be half a wave length. It is not difficult to prove that the distance between two consecutive nodes along the wire will be less than half a wave length, the difference for copper wire about 1 cm. diameter being, in the case of waves whose wave lengths lie between 2 m. and 10 m., a quantity, whose ratio to the half wave length is of the order 10^{-4}, a difference too small to affect the results of observation.

When the wire is terminated, it appears from the theory, § 76, etc. that, owing to the radiation from the free end, the wave fronts are not planes perpendicular to the wire all along it, but gradually approximate to planes as the free end is receded from, the distance of the first node along the wire from the free end being considerably less than a quarter wave length, and the distance between the first and second nodes along the wire being less than half a wave length, the ratio of this difference to the half wave length being a quantity lying between 10^{-2} and 10^{-3}. It follows that the velocity of propagation of electric waves as determined from observations on wires, assuming the distance between consecutive nodes along the wire to be half a wave length, will be very nearly the velocity of propagation of waves in the surrounding medium.

Two methods of comparing the wave lengths along a wire with those in air have been devised by Hertz*. The first method consisted in observing the effect of interference between the waves proceeding from an oscillator and the waves set up in a straight wire by this oscillator; the second method consisted in observing the loops and nodes of the stationary waves set up by reflexion from a plane mirror. Both sets of experiments were made in the same room, though under somewhat different conditions. Two resonators were used, one of which was a circular resonator of diameter 70 cm., and both resonators were in tune with the oscillator. The distance between two consecutive nodes on a straight wire, detected by the circular resonator, Hertz had already found to be 2·8 m., which is in agreement with the theory above, the theoretical value of the fundamental wave length of this resonator being 5·56 m. In the mirror experiment the primary oscillator was at a height of 2·5 m. above the floor, and the resonator was moved along at this height. The first loop occurred at a distance of 1·72 m. from the mirror, the first node at a distance somewhere between 4·1 m. and 4·15 m.; farther from the mirror the phenomena had become so feeble that the position of the second loop could not be determined with any degree of accuracy, all that could be asserted was that it was at a distance somewhere between 6 m. and 7·5 m. from the mirror. It is evident from theory that these results are not those, which would be observed if the phenomena were those of reflexion at a plane mirror, for in that case the distance of the first node from the mirror would be somewhere about twice the distance of the first loop from it, even allowing for the limited extent of the mirror. Disturbing causes, of which the effects are considerable, are present. The possible disturbances are those due to reflexion from the floor and the walls of the room, and

* *Electric Waves*, pp. 107, 124.

those due to the presence of the oscillations in the resonator, which belong to its overtones. The resonator being in tune with the oscillator, the amplitude of the oscillations belonging to its overtones will be very small compared with that of the oscillations, which belong to its fundamental tone, so that, in this case, the effect of these other oscillations in the resonator may be neglected. If it be assumed that the floor is a perfect conductor and that the effect of the walls is negligible, the condition for a well marked loop or node is that the electric force at the gap of the resonator, due to reflexion at the floor and at the mirror, is in the same or opposite phase as the electric force in the gap when the mirror and floor are supposed to be away. Denoting by x and y the distances of the resonator from the mirror and from the floor, these distances being measured parallel to the floor and to the mirror respectively, the above condition is represented approximately by

$$\frac{\sin 2\kappa x}{x} - \frac{\sin 2\kappa r}{r} + \frac{\sin 2\kappa y}{y} = 0,$$

where $\kappa = 2\pi/\lambda$, λ is the wave length and $r^2 = x^2 + y^2$. The number of real roots of this equation in x depends on the distance y of the resonator from the floor; if y is a multiple of half the wave length, $\sin 2\kappa y$ vanishes, and there is an infinite number of real roots; in any other case the number of real roots is finite, as the range of fluctuation of $\dfrac{\sin 2\kappa x}{x} - \dfrac{\sin 2\kappa r}{r}$ diminishes as x increases.

Thus, in general, there will only be a limited number of well marked loops and nodes. In the particular case under consideration the value of $\dfrac{\sin 2\kappa y}{y}$ is very nearly 1/4, all distances being supposed expressed in metres, so that there will be, assuming the floor a perfect conductor, no well marked loop or node, the first loop excepted. Further, the least root x_0 of the above equation is considerably greater than a quarter wave length, and therefore the true wave length in air will be much less than four times the distance of the first loop from the mirror. If the floor is not a perfect conductor, the above equation will be replaced by the equation

$$\frac{\sin 2\kappa x}{x} - \epsilon \frac{\sin 2\kappa r}{r} + \epsilon \frac{\sin 2\kappa y}{y} = 0,$$

where ϵ is some quantity less than unity, and the results will be of the same kind as those described above; the distance of the first loop from the mirror will be greater than a quarter wave length, and a loop or node at some distance from the mirror will not be well marked.

These effects are exactly those which Hertz' results shew to have been present in his experiments, and, it being remembered that the effect of reflexion from a wall would produce the same kind of effects, it can be safely concluded that there was considerable reflexion from the floor or walls of the room, most probably from the floor.

In the interference experiment the primary was differently arranged, its height above the floor being now 1·5 m. The total effect on the resonator in any position is made up of the effect due to the waves coming directly to it from the oscillator, the waves coming from the oscillator, which are reflected to it from the floor and the walls, the waves, which are radiated out from the resonator and are reflected back to it by the floor and the walls, and the waves along the wire. If there were no wire, the waves from the oscillator which arrive at the resonator after reflexion at the walls or at the floor would interfere with the waves coming direct from the oscillator differently at different distances, for the ratio of the amplitudes of these waves and their difference of phase are different at different distances from the oscillator. If then, there being no wire, the resonator were moved in a direction parallel to the floor and away from the oscillator, the sparks would at first diminish in intensity and afterwards increase to a maximum, after which they would again diminish, the successive maxima diminishing and the distance apart of two consecutive maxima tending to, but being always greater than, half a wave length. If there were no reflexion from the walls or floor of the room, the sparks would diminish continuously as the resonator is moved away from the oscillator. When there is a wire, there will be along it waves of different wave lengths, as has been shewn experimentally by Sarasin and de la Rive. Stationary waves will be formed along the wire, whether it has a free end or is led to a conductor, and the electric force perpendicular to the wire will vary along it. The amplitude of the waves along the wire, which have the same wave length as an overtone of the resonator, will not necessarily be negligible in comparison with the amplitude of the waves along the wire, which have the same wave length as the fundamental tone of the resonator. The effect on the resonator of the waves along the wire will therefore be a periodic function of the distance measured parallel to the wire. The nature of the effect on the resonator of the interference of the waves along the wire with the waves coming directly from the resonator and those coming from the walls and the floor of the room will then be of the character observed by Hertz; if there were no reflexion from the walls or floor, or if there were only simple harmonic waves along the wire, the interference would not have been of the kind observed; to produce it, there must have been waves along the wire with the same wave length as one or more of the overtones of the resonator, as well as reflexion from the walls or floor of the room.

These experiments of Hertz taken together shew indirectly that the overtones of a resonator are, under the proper conditions, excited. Evidence for their existence can also be obtained from the experiments of Sarasin and de la Rive. They found in their mirror experiments that there was no position in which the sparks totally disappeared. If only the fundamental tone of the resonator had been excited, there ought theoretically to be no sparks at a distance from the mirror, which is a multiple of half the fundamental wave length of the resonator, and if the wave lengths corresponding to the overtones were integral submultiples of the fundamental wave length of the resonator, there would be no sparks, due to the presence of the overtones, at such points. It has been shewn, § 71, that, if λ_0, λ_1 are the wave lengths of the fundamental tone and of the first overtone,

$\lambda_1 = \cdot 28\lambda_0$, whence $7\lambda_1/4 = \cdot 49\lambda_0$, and therefore the first node from the mirror for the fundamental tone is very near to a loop for the first overtone. If the fundamental wave length of the resonator is nearly the same, or less than, the wave length of the oscillations sent out by the oscillator, the amplitude of the oscillations belonging to any of the overtones of the resonator will be small compared with that of the oscillations belonging to the fundamental tone, so that, in such cases, the observed positions of the nodes and the loops will be very approximately the same as if the overtones of the resonator were not excited; the only appreciable effect of the overtones will be that there will be no position in which sparks are entirely absent. If, however, the fundamental wave length of the resonator is greater than the wave length of the oscillations sent out by the oscillator, the amplitude of the oscillations belonging to the first overtone of the resonator may not be small compared with the amplitude of the oscillations belonging to the fundamental tone, as the ratio of their amplitudes is very approximately

$$\left\{ \left(\frac{1}{\lambda^2} - \frac{1}{\lambda_0^2} \right) \middle/ \left(\frac{1}{\lambda_1^2} - \frac{1}{\lambda^2} \right) \right\}^{\frac{1}{2}},$$

where λ is the wave length of the oscillations of the oscillator, and, in such cases, the observed positions of the nodes and the loops would not be the same as if the overtones were not excited. This is probably the explanation of the fact that Sarasin and de la Rive in their third series of experiments did not obtain good results in the case of the resonator of diameter 1 m., as its fundamental tone has a wave length considerably greater than that of the oscillations of the oscillator, and the amplitude of the oscillations belonging to the first overtone of this resonator is approximately, in the case of the oscillator used, one-fifth of the amplitude of the oscillations belonging to the fundamental tone, so that, in this case, there would be a displacement of the observed positions of the nodes and the loops from the positions of the nodes and the loops belonging to the fundamental tone, which, though small, would be appreciable. Another curious effect, which they observed, was that the intensity of the minimum sparks in the resonator was greater than the intensity of the sparks, when the mirror was away*. This can be explained as follows: when there is no mirror, the almost dead beat oscillations of the oscillator produce an effect on the resonator similar to that which would be produced by a single pulse. When the mirror is present, the waves from the oscillator, after striking on the resonator, are reflected back from the mirror, as are also the waves emitted from the resonator, and these latter have a very small rate of decay. If the resonator is at a distance from the mirror which would be a loop for one of its overtones, the oscillations belonging to this overtone are continually reinforced, and after a time their amplitude will become great compared with that of the amplitude of the oscillations excited by the oscillator when there is no mirror. As the rate of the radiation of energy from a resonator varies inversely as the fourth power of the wave length, the amplitude of the oscillations belonging to an overtone will be much less in comparison with the amplitude of the oscillations belonging to the fundamental tone than it is when there is no mirror, so that the results stated above, as to the effect of the overtones on the positions of the loops and the nodes, will not be affected. The effect of the overtones will be that observed by Sarasin and de la Rive; there will be no position in which the sparks totally disappear in front of a reflector, and the intensity of the sparks in the resonator, when it is in a position in which the intensity of the sparks is a minimum, will be greater than the intensity of the sparks produced in the resonator, when there is no reflector.

---

* The effect of even slight reflexion, such as that from a wall, in increasing the intensity of the sparks in a resonator had already been observed by Hertz.

APPENDIX A.

THE RELATION OF THEORETICAL TO EXPERIMENTAL PHYSICS.

ALL observations consist in the comparison of motions, the thing observed being always a change of position of something as, for example, of a needle, a spot of light or the hand of a watch. The function of theoretical physics is to give a consistent representation of these changes, and this representation has to be made in terms of certain general conceptions such as space. Now space is conceived of as possessing the property of extension only, and it is postulated of it that it is possible to assign definite positions (points) in space, to draw uniquely a line from any one point to any other, which shall have the same direction throughout, and to draw from any point uniquely a line which shall throughout have a definite assigned direction. Thus space itself is not conceived of as moving, but as being such that motion can take place in it. The possibility of the motion of a point in space involves that of its not being in motion, so that absolute position in space is a necessity of thought.

If AB and CD be two parallel straight lines and points P and Q be supposed to move along AB and CD respectively so that the straight line PQ moves from the position AC to the position BD, the points P and Q are said to describe AB and CD in the same time, and further, if the straight line PQ in all intermediate positions which it occupies passes through the intersection of AC and BD, the rates of description of AB and

CD by *P* and *Q* are said to be always in the constant ratio *AB* to *CD*.

In this way rates of description of paths, that is velocities, can be compared; hence arises the idea of a constant velocity, and time is conceived of as measured mathematically by a point which describes a definite path, as for example a circle, with a constant velocity. Mathematical time is not necessarily the same as the time of actual experience, no amount of experience can ever prove that they are the same or that they are different. What can be asserted is that there are actual timekeepers which measure time in such a way that within the range of actual experience its properties are indistinguishable from those which are assigned to the time of mathematical thought.

Similarly, it cannot be asserted of any velocity of actual experience that it is an absolute velocity; what can be is that within the range of the particular experience under consideration its behaviour is indistinguishable from that of a mathematical absolute velocity or the difference of two such velocities. There can be no actual experience of the absolute space, velocity or time of mathematical thought, but it does not thence follow as some writers have held that we can have no knowledge of them. These conceptions have been evolved by the mind to suit its mode of action which impels it to represent as far as it can the phenomena of actual experience in terms of things which can be thought of as forming parts of the whole conceptual system and at the same time as existing independently.

Anything capable of actual motion was originally termed matter, and the motions observed were discussed on the assumption that they could be represented by a moving point or points. If the laws of motion according to which such a point moved were known, the complete history of the motion could be traced, the position of the point at any instant being then expressible in terms of its position and velocity at a definite time. In attempting to formulate the laws of motion of moving points, whose motions should represent actual motions, it was natural to make use of the ideas already arrived at concerning matter. Matter was thought of as indestructible and measurable, the quantity of matter in a body being independent of its position and unalterable with the time. Further, motion could be communicated to a portion of matter, the simplest case being that where motion is communicated from one moving body to another, the communication of motion being supposed to take place instantaneously. At first the quantity of matter in a body was determined by its weight; it was only when observation had shewn that the weight of a body was not independent of its position that it became clear that quantity of matter was not identical with weight. Matter was recognized as being of different kinds, and bodies, the material of which was the same, could be compared in respect of quantity by the volumes they occupied; comparison between bodies composed of different kinds of matter would be meaningless except in respect of some property which all kinds of matter possessed in common. Their common property being that they can move, the comparison between them must be made in respect of the communication of motion from one body to another, and in the theoretical comparison each body is supposed to be capable of being represented by a moving point.

Observation shews that the circumstances of the motion of two bodies after the communication of motion between them depend on the directions of their motions before, so that the simplest case is that in which the motions of the two bodies are such that they can be represented by two points moving in the same straight line, the object of discovery being some quantity which is unaltered. It is then found that there is a linear function of the velocities which remains constant, that when the two bodies are of the same material the coefficients of their velocities in this expression are in the same ratio as their volumes, and also that, if communication of motion takes place between a body A_1 and a body B of different material and between a body A_2 of the same material as A_1 and the body B, if the coefficient of the velocity belonging to B be taken to be the same in the two cases, the ratio of the coefficients of the velocities belonging to A_1 and A_2 is the same as when communication of motion takes place between A_1 and A_2. Further the knowledge so far obtained of these coefficients makes it natural to expect that, when communication of motion takes place between two bodies which are not moving in the same straight line, the same linear function of the component velocities in any direction remains constant, and this is found to be the case. It follows, that all material bodies can be compared in respect of the communication of motion from one to another; the coefficient belonging to any one body being kept constant, the others remain constant, having always the same ratio to the standard one. The linear function of the velocities in any direction which remains constant is termed the momentum* of the bodies in that direction, and the part of it which involves only one of the velocities, the momentum of the body possessing that velocity, these momenta being compounded like velocities. The ratio of the coefficients belonging to any two bodies is termed the mass-ratio of these two bodies, and, it being agreed that the coefficient belonging to a particular body is unity, this body is said to be of unit mass and the coefficient belonging to any other body the measure of its mass. When communication of motion takes place between two bodies the effect produced in any direction is measured by the change of momentum of the body in that direction. Observation shews that in many cases change takes place in the motion of a body without its being immediately assignable to communication with another moving body; the change which takes place is still measured in the same way, and the cause producing this change is termed force, the measure of the force being that of the change which it produces. When the changes are such that they must be considered as taking place gradually and not instantaneously, the corresponding laws are obtained by considering the gradual change to be the limit of a great number of changes taking place instantaneously at successive small intervals of time. Change of momentum is then replaced by time rate of change of momentum and force by continuous force. The preceding constitutes the Newtonian scheme for the representation of the phenomena of bodies in motion and in it the idea of momentum is fundamental. The idea of mass is

* By some writers it has been termed the "quantity of motion."

arrived at subsequently, and it is a result of experience that the mass-ratio of two bodies of the same material is that of their volumes, just as it is a result of further experience that the masses of bodies can at the same place be compared by their weights.

Instead of taking momentum as the fundamental idea another measure of the quantity of motion might have been chosen. That adopted by Huygens was the quantity now known as energy, and with it as the fundamental idea a complete scheme for the representation of the phenomena can be developed. So far as mechanics is concerned either scheme is equally convenient; it is when dynamical methods come to be applied to the representation of other physical phenomena that the advantages of the second scheme become apparent. For some time physical phenomena such as those of heat and electricity were ascribed to the existence of special substances which were different from matter, though measurable. Gradually, however, attention was directed to the fact that motion could be converted into heat and heat into motion. It was then discovered that quantity of heat could be measured in terms of motion, the energy of the corresponding motion being a measure of the quantity of heat. The principle of the conservation of energy, which was previously known to hold for mechanical phenomena, was thus extended to thermal phenomena and then to all physical phenomena. The natural inference was made that all physical phenomena are modes of motion, though in many cases these motions could not be directly observed. The fact that some of the motions can not be directly observed makes it convenient that the measure of quantity of motion should be independent of its direction, the simplest such measure being the energy of the motion. Further, it has to be recognized that all these motions are not necessarily motions of matter, the term matter being now restricted to that which possesses molecular structure. Motions of other kinds which may not be capable of being represented by moving points must be contemplated, and the idea of mass must not be associated with these motions as this idea involves the possibility of representing them by moving points. In order then to apply dynamical methods to all physical phenomena the laws of dynamics must be expressed in terms of that measure of the quantity of motion which is termed energy and in a form independent of the idea of mass, these laws not being inconsistent with the laws of motion of material bodies. Now these conditions are fulfilled by that form of the Lagrangian method which is usually known as the principle of Least Action and which expresses the fact that the system moves so as to expend as much energy as it can. The object of discovery in any case is the Lagrangian function and this function has to be constructed from the results of experience. The presence of motions other than those which can be represented by moving points being contemplated, there may be degrees of freedom which cannot be specified by the coordinates of moving points, and the coefficients of the squares of the velocities corresponding to these coordinates in the Lagrangian function must be determined from the results of experience just as has been done in the case of the coefficients of the squares of the velocities of the moving points which represent the motions of material bodies. On this view all forces are to be regarded as forces of motion, motion being the change underlying all the changes which constitute physical phenomena. This being so, it seems natural to inquire whether the Lagrangian functions which are constructed from the results of experience can be derived from the Lagrangian function which would involve all the degrees of freedom belonging to the motions which constitute physical phenomena and not those only which are directly observed. This is what has been done in Chapter v. above, where the forms of the modified Lagrangian function which arise from the elimination of a number of the coordinates specifying degrees of freedom have been discussed and where, in particular, it has been shewn that the modified Lagrangian function, which occurs in a great number of cases in the form of the difference of two functions—one a homogeneous quadratic function of the time rates of variation of the coordinates which specify the observed degrees of freedom and the other a function of these coordinates—arises from the original Lagrangian function when the latter involves the coordinates specifying the unobserved or concealed degrees of freedom in a particular way, and that the function occurring in the modified Lagrangian function which does not involve time rates of variation of the coordinates specifying observed degrees of freedom and usually termed the potential energy is the energy of the concealed motions. Now the addition to the Lagrangian function of a constant or of a linear function of the time rates of variation of the coordinates in evidence in it, the coefficients in this linear function being constants, does not alter the equations of motion; thus the presence of degrees of freedom which are involved in this manner in the Lagrangian function does not affect the motions corresponding to the other degrees of freedom. Further, if there are degrees of freedom whose coordinates are involved in the Lagrangian function in such a way that it differs from the Lagrangian function, which would exist if they were absent, by the addition of parts which are (1) approximately a constant and (2) a linear function of the velocities belonging to the other degrees of freedom, the coefficients in this function being approximately constant for the range throughout which the motions belonging to these latter degrees of freedom are being considered, these motions will not be appreciably altered.

Thus, in the case of the motion of material bodies, the equations of motion will be approximately the same whether the motions be measured relatively to a body A, or to a body B which has a small motion relatively to A, provided that the time for which the motions are being considered is sufficiently short. The laws of motion of material bodies were first arrived at from observation of moving bodies in the immediate vicinity of some place on the earth's surface, the motions being measured relatively to this place. When motions of bodies near the earth's surface, these motions lasting a considerable time as in the case of bodies falling from a great height, were made the subject of observation, it was found to be convenient, in order to obtain a simple representation of the motions, to measure the motions relatively to axes supposed to be drawn through the centre of the earth, one of them being the axis about which the earth rotates relatively to the stars and the other two fixed in direction relatively to the stars.

Similarly in the case of the motions of the planets it was found convenient to measure the motions relatively to axes supposed drawn through the sun and fixed in direction relatively to the stars. If the first class of motions be measured relatively to the axes supposed drawn through the centre of the earth in the second case, the resulting motion relatively to the axes first drawn will be approximately the same, and similarly, if the motions of both these classes be measured relatively to the axes drawn through the sun in the third case, the resulting motions relatively to the sets of axes first chosen in the respective cases will be approximately the same, the differences in the several cases being experimentally inappreciable.

In every case of the motion of material bodies the laws of motion must be the same, and if the natural assumption, that of Newton, be made, that the force between any two material bodies depends only on their relative position, there will be a set of bodies, possibly imaginary, there being no body not included in the set which possesses only a uniform motion of translation relatively to the set, such that if motions are measured relatively to them, the laws of motion will be accurately true, whilst they will only be approximately true if the motions be measured relatively to any other body not included in the set. In the case of motions generally there will be a body, possibly imaginary, such that if all motions be supposed to be measured relatively to it the laws of dynamics will be accurately true, while they are only approximately true when the motions are measured relatively to any other body. In the theory of electrodynamics Faraday's laws were first arrived at from observation of circuits conveying electric currents, the positions and motions of these circuits being measured relatively to the place on the earth's surface at which the observations were being made, and, when Maxwell investigates the state of affairs in the intervening medium by exploring it by means of the secondary circuit, this circuit is in each position it occupies supposed to be fixed relatively to the place of observation. When the propagation of electrical effects across distances comparable with those of Astronomy comes to be discussed, the motions are measured relatively to the axes of reference of Astronomical investigations, and the secondary circuit of exploration must now in each position it occupies be supposed to be fixed relatively to these axes. The laws of electrodynamics arrived at from the previous experiences are assumed to be still true, an assumption justified by observation, and if the motions in these previous experiences be measured relatively to the Astronomical axes the differences between the results so obtained and those previously obtained are so small as to be incapable of detection by these observations. There will then in the case of electrodynamics be some body such that, if all motions be measured relatively to it, the laws of electrodynamics will be accurately true, and this case differs from that of mechanics, inasmuch as the laws of electrodynamics would not be accurately true if the motions were measured relatively to a body which had a uniform motion of translation relatively to the body of reference. What is to be understood by the term "axes fixed in space" is a set of axes which are such that if all motions were measured relatively to them the laws of electrodynamics would be accurately true. In the comparison of theory and experiment the set of axes to be used instead of those of theory is that set which is most convenient and the use of which will lead to a sufficiently accurate representation of the phenomena under discussion.

APPENDIX B.

CONTINUOUS MEDIA.

A CONTINUOUS medium may be defined as being a medium, which is such that in any region of space occupied by it no point can be found which is not also in the medium. The idea of continuity present in this definition is identical with that of the geometrical continuity of space. Rigid bodies and elastic solids, liquids and compressible fluids have, in order to subject them to mathematical treatment, been identified with continuous media, and it is important to inquire what the process known as "treating them as continuous" is equivalent to, so as to find out how far the mathematical treatment, which is applicable to them, can be applied to a continuous medium such as the aether must be postulated to be. When the motions of some specified physical body are mathematically investigated, this body must be replaced in imagination by some object which can be completely defined. It is usual to assert that all material media possess atomic or molecular structure, and this being assumed to be so, it is first necessary to try to state in what way a molecule or atom can be represented so as to be capable of mathematical treatment. A molecule or an atom may be regarded as having a certain quantity of energy associated with it, this energy being expressed in terms of a number of coordinates which are such that all possible variations of the energy can be taken account of by varying these coordinates. A complete representation of this kind has still to be obtained, but it would appear that in a large number of cases of motions of material bodies the

following representation is sufficiently accurate. The position
of the molecule or atom in space at any time may be supposed
to be identified with that of some point whose coordinates
are x, y, z, the axes of reference being the fixed axes of
theoretical dynamics. The energy associated with the molecule
or atom may be supposed to consist of three parts, a part
depending on the velocity of the point which determines the
position of the molecule or atom and which is $\frac{1}{2}m\,(u^2 + v^2 + w^2)$,
where u, v, w are the component velocities of the point and m
is a mass coefficient, a part which is a function of x, y, z and
of the coordinates of the points which determine the positions
of any other molecules or atoms which may be present, and a
part which is a function of other coordinates not in evidence ;
this latter part must be supposed to be invariable as must also
the coefficients of the second part, it being conceivable that
these coefficients may be functions of the other coordinates
which are not in evidence.

A material medium may be regarded as an aggregate of
molecules or atoms, and the mathematical theory of such
material media can then be developed from the following
assumptions :—

The matter which occupies any finite volume of space is
regarded as being constituted by an aggregate of molecules.

The motion of each molecule can be represented by that
of a moving point.

The effect of all other motions, whether they are motions
of the molecules which cannot be represented by moving points
or motions which do not belong to molecules, can be represented
by a potential energy function or by a system of forces associated
with each molecule, this function or system of forces depending
only on the positions of the molecules ; and this includes the
case in which the effect of these other motions, instead of being
immediately represented by a system of forces associated with
each molecule, is represented by restrictions on the possible
motions of the molecules.

The Lagrangian method is applicable to the motions of the system.

The kinetic energy of a molecule is then represented by $\frac{1}{2}m\,(u^2 + v^2 + w^2)$, where u, v, w are the component velocities of the point whose motion represents the motion of the molecule, and m is the mass belonging to the molecule. The kinetic energy of the aggregate is represented by $\sum\limits_{n=1}^{n=N} \frac{1}{2}m_n(u_n^2 + v_n^2 + w_n^2)$, where the suffix n identifies a particular molecule and the summation extends to all the molecules, there being N of them. Similarly, the potential energy of the aggregate is represented by $\sum\limits_{n=1}^{n=N} m_n V_n$, or the work done by the systems of forces, which represent the effect of the other motions, is

$$\sum_{n=1}^{n=N} m_n\,(X_n\delta x_n + Y_n\delta y_n + Z_n\delta z_n)$$

for small arbitrary displacements of the molecules. The application of the Lagrangian method will then lead to $3N$ equations involving in addition to the $3N$ coordinates, which specify the positions of the molecules, as many un-determined multipliers as there are relations specifying restrictions on the possible motions of the molecules, and these undetermined multipliers specify the systems of forces which are equivalent to the restrictions on the possible motions of the molecules. When these $3N$ equations, the equations restricting the motions of the molecules being taken into account, have been solved, the positions and motions of the molecules at any time are expressed in terms of their positions and motions at a particular time, and the complete history of the motions can be traced. In general the number of dependent variables occurring in any one of the equations will be great, and the solution of the equations will be practically impossible, but, if the equations form a set of groups such that the number of dependent variables occurring in the equations of a group is equal to the number of equations in the group, and this number is not too large, the solution of the equations becomes more possible, the simplest case being that in which the groups are exactly alike. The most important case is that in which the discrete analysis sketched above can be replaced by a continuous analysis. If a surface be drawn enclosing a simply-connected space in which there is a number of molecules, the ratio of the mass of the molecules inside this surface to the volume enclosed by it may be termed the average density of the aggregate of molecules inside the surface. Now let such a surface surrounding the point O be contracted so that the volume enclosed by it is diminished, the average density inside the surface will tend towards a limit ρ so long as the least linear dimension of the volume enclosed is not less than a certain length depending on the distance between two adjacent molecules, but will afterwards depart altogether from that limit*. This limit ρ may be termed the density at the point O, and will be a function of the position of O, such that $\iiint \rho \, dx \, dy \, dz$ taken throughout any volume differs from the sum of the masses of the molecules in that volume by a mass which is less than some standard small mass. By an exactly similar process of reasoning it may be shewn that the expression $\sum_{n=1}^{n=N} \frac{1}{2} m_n (u_n^2 + v_n^2 + w_n^2)$ for the kinetic energy may be replaced by the expression

$$\iiint \tfrac{1}{2} \rho \, (u^2 + v^2 + w^2) \, dx \, dy \, dz,$$

where the integral is taken throughout the volume occupied by the aggregate of molecules, this integral differing from the sum by a quantity which is less than some standard small kinetic energy. These remarks also apply to the potential energy function or the systems of forces associated with the molecules, which represent the effect of all the other motions which may exist. In the discrete analysis the suffixes n serve to identify the molecules. In order then that the equations of motion obtained by the application of the Lagrangian method in the continuous analysis may legitimately replace the equations of motion in the discrete analysis, the assumption must be made that throughout its motion any point x, y, z is always identified with the same bit of matter, and, when the system under consideration is being regarded as an aggregate of molecules,

* The limit is not actually attained.

this requires the following condition to be satisfied. If a surface be drawn passing through a number of points which specify the positions of molecules, and this surface move with the molecules, no molecule which is initially on one side of this surface ever comes to be on the other side of it. This may be expressed by saying that the order of arrangement of the molecules does not change. When the above conditions are satisfied, the Lagrangian method can be applied, and in this way a group of equations is derived, this group being typical of all the equations of motion of the molecules. The media for which the Lagrangian function is of the kind specified above, and to which the operations of mathematical analysis are applicable, will then consist of aggregates of molecules, for which the order of the arrangement of the molecules does not change, and which are such that the distance between any two adjacent molecules is very small.

The "rigid body" of mathematical theory satisfies all the above conditions, for the distance between any two molecules of the aggregate which constitutes the body is assumed to be invariable, and therefore the order of arrangement of the molecules does not change. When the aggregate of molecules is such that the distance between any two neighbouring molecules can vary, but only to a small extent, and the distance between any two adjacent molecules is very small, the aggregate constitutes what is termed an "elastic solid." When the system of forces or the potential energy function associated with each molecule is known, it being assumed that the order of arrangement of the molecules does not alter, the Lagrangian method can be applied, and the resulting equations will determine the history of the changes which take place. The system of forces or the potential energy function associated with each molecule will consist of two parts, of these one is determined by the changes in the distances between the molecules, the other must be supposed to be given as in the case of the rigid body. The first object of discovery is then the system of forces or the potential energy function associated with each molecule, which arises from the change in the distances between the molecules. Two methods of effecting

10—2

this have been proposed. The first of these, due to Navier and developed by Poisson, Cauchy and St Venant, consists in assuming that the effect of the molecules on each other can be represented by forces between them, the force between any two depending only on their distance apart, and the direction of this force being that of the line joining them. The processes by which the stresses are then expressed in terms of the strains involve the assumptions that the order of arrangement of the molecules does not change and that the constitution of a molecule remains the same, that is, that during the displacements the molecules are not broken up and then formed into new molecules. These stresses possess a work function, and, when the medium is isotropic, this function involves only one constant. The other method of obtaining the stresses is due to Green and consists in assuming that the stresses possess a work function, this function being expressible in terms of the strains. The application of the Lagrangian method then leads to equations which are sufficient for the determination of all the circumstances of the changes which take place ; but in order that the work function should be of the form assumed and that the Lagrangian method should be applicable, it is necessary to assume that the order of arrangement of the molecules does not change and that the molecules do not break up during the displacements. When the medium is isotropic, Green's work function involves two constants, and the distinction between his theory and that of Navier is that on Green's theory the intermolecular forces cannot be represented by forces between pairs of molecules, the magnitude of the force between a pair depending only on the distance between them. It would appear from the investigations of Chapter VIII. above that, if intermolecular forces be assumed to be of electrical origin, the force which represents the effect of a molecule on any other molecule is not necessarily a function of the distance between the two, but the forces, from the way in which they are derived, would possess a work function, so that a work function of the form assumed by Green is possible. The question, whether Navier's theory or Green's furnishes the best representation of the phenomena, has been much discussed, and attempts have been made to decide the question by an appeal to experiment, but the evidence from statical experiments cannot be accepted for a reason which will appear later. One of the strongest arguments in favour of Green's form of the work function is that it is the most general one, which permits of the application of the methods of mathematical analysis, and is such that the parts of the medium are in a state of relative rest when there is no external system of forces.

In the above the idea, that a molecule is itself composite, has been introduced, and it is important to inquire what the restrictions are under which the aggregates of atoms, each molecule being now thought of as consisting of an aggregate of atoms, can, in the application of the methods of mathematical analysis, be replaced by the aggregate of molecules. If each one of the atoms in the aggregate of atoms which constitutes any molecule is describing an orbit, which is periodic relative to some point x, y, z, this point defining the position of the molecule relative to the other molecules, the position of any atom at any time is specified by the coordinates $x + \xi$, $y + \eta$, $z + \zeta$. The equations of motion of the whole system will then be contained in the variational equation

$$\Sigma\Sigma m\,(\ddot{x} + \ddot{\xi})\,\delta x + \ldots + \ldots + \Sigma\Sigma m\,(\ddot{x} + \ddot{\xi})\,\delta\xi + \ldots + \ldots$$
$$= \Sigma\Sigma m X \delta x + \ldots + \ldots + \Sigma\Sigma m \Xi \delta\xi + \ldots + \ldots,$$

where m is the mass of an atom, and the summations are taken for all the atoms in each molecule and then for all the molecules. If from these equations all the coordinates x, y, z be eliminated, there will result equations from which the differences $\xi_1 - \xi_2$, $\eta_1 - \eta_2$, $\zeta_1 - \zeta_2$, of the coordinates ξ, η, ζ can be determined. The integrals of these equations, the orbits of the atoms, relative to the points specified by the coordinates x, y, z, being all periodic, will be of the form

$$\xi_1 - \xi_2 = \Sigma A_s \cos\left(\frac{2\pi t}{T_s} + \alpha_s\right);$$

whence it may be assumed that

$$\xi = \Sigma \bar{\xi}_s \cos\left(\frac{2\pi t}{T_s} + \alpha_s\right).$$

Now the coordinates x, y, z can always be chosen so that they do not involve parts which are periodic in any of the periods T_s, and when the above expressions for ξ, η, ζ are supposed to be substituted in the variational equation, it will take the form

$$\Sigma M\ddot{x}\,\delta x + \ldots + \ldots = \Sigma M X'\,\delta x + \ldots + \ldots ,$$

where M is the mass belonging to a molecule. In this equation X', Y', Z' will represent the systems of forces associated with each molecule, provided that neither the part of them, which arises from intermolecular action, nor the part, which does not arise from such action, involves a term which is periodic in any of the periods T_s. The aggregates of atoms can therefore, in the application to them of the methods of mathematical analysis, be replaced by the aggregate of molecules when the following conditions are satisfied.

Each atom in a molecule describes an orbit which, relative to some point defining the position of the molecule to which it belongs, is periodic*, and the system of forces associated with each molecule, which does not arise from action between the molecules, does not contain a part which is periodic in any of the periods belonging to the orbits of the atoms. Further, when these conditions are satisfied, the part of the inter-molecular forces, which is effective in respect of the positions of the molecules, is the part which is independent of the periods belonging to the orbits of the atoms.

When forces, which represent the effect of moving systems other than the aggregates of atoms, act on the atoms, the course of events can be represented as follows; if no part of these external forces is periodic in any of the periods of the orbits of the atoms, they are equilibrated by the motional forces of the molecules and the part of the intermolecular forces which is not periodic in any of the periods of the orbits of the atoms. Their effect, in the first instance, will be to change the relative distances of the molecules; this change produces a change in the part of the intermolecular forces which is periodic in the periods of the orbits of the atoms, and to balance this the

---

* Such an orbit is not necessarily closed.

dimensions of the orbits of the atoms are changed. The extent to which the dimensions of the orbits of the atoms can be changed is limited by the condition that the displaced orbits must be stable, and the changes in the orbits of all the atoms belonging to a molecule will be related in such a way that there is no radiation of energy from it.

When there are external forces which are periodic in the periods of the orbits of the atoms, an analysis which only takes account of an aggregate of atoms will not be applicable; a finer analysis, which shall take account of the atoms whose orbits have among their periods the periods of the external forces, must be used.

Again, the external forces may be such that an analysis which takes account of each molecule is unnecessary, and, a coarser analysis which only takes account of compound molecules, each compound molecule being made up of a number of molecules, is sufficient.

In every case the range, throughout which the continuous analysis will be applicable, is determined by the conditions that the units, whose motions are taken into account, whether these units are compound molecules, molecules, or atoms, do not change their order of arrangement, and that, if they are composite, they do not break up. The work done by the external forces is all accounted for by the changes in the positions and in the motions of the units, and therefore a continuous analysis is only strictly applicable to an aggregate of molecules or atoms, when the changes which take place are such that no energy is gained or lost in the form of heat or any other form which is not taken account of by the changes in the positions and motions of the molecules or atoms.

The possibility of applying continuous analysis to an elastic solid therefore requires that the changes which take place in it shall take place adiabatically. When the elastic solid is executing small vibratory motions, this is probably true, and in this case continuous analysis can be applied to it. When an elastic solid passes from one state of strain to another by the

application of external forces, the changes taking place in such
a way that a continuous analysis is applicable, the intrinsic
energy of the molecules will be altered, and, if there are no
other influences present, the temperature of the body will be
different in the two states. In any actual case the body
is surrounded by some other medium, and the initial state
of the body is one of equilibrium relative to its surroundings;
the state arrived at after the changes have taken place adiabatically will not be one of equilibrium relative to its surroundings, to attain a state of equilibrium a transference of energy
will take place between the body and the surrounding medium.
This transference of energy, supposed to take place in such
a way that a continuous analysis is still applicable, ought to be
taken account of in the equations which give the history of the
changes, and the result would be the introduction into these
equations of a system of forces which would represent the effect
of the surrounding medium.   For example, when the Young's
modulus of any material is determined by stretching a wire
made of this material by a weight attached to it, the work done
by the weight is not all used up in stretching the wire, some of
it is used in altering the condition of the surrounding medium,
and therefore the Young's modulus of the material so determined
will differ from the true one.

When an aggregate of molecules is such that the molecules
can move freely and the distance between any two adjacent
molecules is very small, the aggregate forms what is termed
a fluid.   As in the case of an elastic solid a continuous
analysis will only be applicable when the molecules move in
such a way that their order of arrangement does not alter and
no energy is gained or lost in the form of heat. Since the order
of arrangement of the molecules does not alter, the same molecules will always be at the boundary, no molecule which is not
originally at the boundary will ever come there, and no molecule
which is originally at the boundary will ever cease to be there.
Further, if at a definite time t_0 a certain number of molecules
occupy an element of volume at the point x_0, y_0, z_0, this point
defining, for the purposes of continuous analysis, the position of
these molecules, the same molecules will at a time t occupy an

element of volume at the point x, y, z, where this point defines the position of the same molecules at the time t, and therefore, the mass belonging to the molecules being unalterable, the relation

$$\rho \, \frac{\partial (x, y, z)}{\partial (x_0, y_0, z_0)} = \rho_0$$

will be satisfied at each point of the space occupied by the fluid. This relation expresses the correspondence which exists between the spaces occupied by the fluid at two different times, in consequence of the restriction on the possible motions of the molecules. If the forces which represent the effect of other systems on the aggregate of molecules possess a work function, the Lagrangian function of the motions will have the form

$$\iiint \left[\tfrac{1}{2}\rho \, (\dot{x}^2 + \dot{y}^2 + \dot{z}^2) - \rho V \right] dx \, dy \, dz.$$

For a liquid $\rho = \rho_0$, ρ_0 being a constant for a homogeneous liquid and a known function of x_0, y_0, z_0, for a heterogeneous liquid. The relation, which expresses the restriction on the possible motions of the molecules of the liquid, is

$$\frac{\partial (x, y, z)}{\partial (x_0, y_0, z_0)} = 1,$$

and the application of the Lagrangian method gives

$$\delta \int_{t_0}^{t_1} \iiint \left[\tfrac{1}{2}\rho \, (\dot{x}^2 + \dot{y}^2 + \dot{z}^2) - \rho V \right] dx \, dy \, dz \, dt = 0,$$

subject to the above relation, that is

$$\delta \int_{t_0}^{t_1} \iiint \left[\tfrac{1}{2}\rho_0 \, (\dot{x}^2 + \dot{y}^2 + \dot{z}^2) - \rho_0 V + \kappa \, \frac{\partial (x, y, z)}{\partial (x_0, y_0, z_0)} \right] dx_0 \, dy_0 \, dz_0 \, dt = 0,$$

where κ is an undetermined function of x, y, z.

This is equivalent to

$$\int_{t_0}^{t_1} \iiint \left[\rho_0 \, (\dot{x}\,\delta\dot{x} + \dot{y}\,\delta\dot{y} + \dot{z}\,\delta\dot{z} - \delta V) + \kappa\delta \, \frac{\partial (x, y, z)}{\partial (x_0, y_0, z_0)} \right].$$
$$. \, dx_0 \, dy_0 \, dz_0 \, dt = 0,$$

that is,

$$\iiint \rho \, \dot{x}\,\delta x \, dx \, dy \, dz \int_{t_0}^{t_1} + \int_{t_0}^{t_1} \iint l\kappa \, \delta x \, dS \, dt + \dots + \dots$$
$$- \int_{t_0}^{t_1} \iiint \left[\rho\ddot{x} + \rho \, \frac{\partial V}{\partial x} + \frac{\partial \kappa}{\partial x} \right] \delta x \, dx \, dy \, dz \, dt + \dots + \dots = 0.$$

The quadruple and the triple integrals must separately vanish; that the quadruple integrals may vanish, the equations

$$\rho\ddot{x} + \rho\frac{\partial V}{\partial x} + \frac{\partial \kappa}{\partial x} = 0,$$

$$\rho\ddot{y} + \rho\frac{\partial V}{\partial y} + \frac{\partial \kappa}{\partial y} = 0,$$

$$\rho\ddot{z} + \rho\frac{\partial V}{\partial z} + \frac{\partial \kappa}{\partial z} = 0,$$

must be satisfied throughout the space occupied by the liquid at any time, and the equations which result from the vanishing of the triple integrals give the dynamical conditions satisfied at the boundary at any time, the form of these conditions depending on the nature of the boundary. The function κ which occurs in these equations represents the internal system of forces which arises from the restrictions on the motions of the molecules and is known as the pressure. There are two kinds of possible motions of a liquid; the motions which can be generated by external conservative forces, and the motions which cannot be so generated. In the first of these the motion is what is termed irrotational and in the second rotational or vortex motion. The possibility of applying continuous analysis to the motions of a liquid involves the result that vortex motion is always associated with the same molecules. Further, every motion of a liquid due to external causes can be represented as the effect of a vortex sheet, whose surface is the boundary of the liquid to which the continuous analysis is applicable. The representation can also be effected in terms of other distributions of vortex motion external to the liquid; for example, when a portion of the liquid is in a state of motion which is not necessarily capable of being treated by a continuous analysis, the effect of this portion on the remainder of the liquid, supposed to be moving in such a way that a continuous analysis is applicable to it, can be represented by a distribution of vorticity throughout the first portion, but it cannot be thence inferred that this vortex motion constitutes a geometrical representation of what is going on in this first portion of the liquid. When an actual liquid is in contact with an actual solid, in passing from the solid to the liquid a transition is made from a place where the molecules are, relatively to each other, approximately immobile, to a place where the molecules approximate to a condition of perfect mobility. The nature of the forces which hold the molecules together will therefore, throughout a certain space in the neighbourhood of the surface of the solid, change their character; in the solid they will be of a nature similar to that of the internal forces in the theoretical elastic solid, and in the liquid, at some distance from the surface, they will be capable of being represented by a pressure; in the portion of the liquid in the immediate neighbourhood of the solid they cannot be so represented. The continuous analysis which is applicable to the motions of the theoretical liquid cannot be applied to this portion, but the effect of the solid and of this portion of the liquid on the portion of the liquid to which the continuous analysis can be applied, can be represented as being due to a vortex sheet whose surface is the boundary of that portion of the liquid to which the analysis is not applicable. If this surface can be taken so near to the solid that the distance of any point in it from the surface of the solid is so small as to be negligible, the condition to be satisfied at the surface of the solid will be that the velocities of the solid and of the liquid normal to this surface are continuous. When a solid is moving through a liquid such as water, which may be taken to be perfectly mobile, the portion of the liquid in the neighbourhood of the solid, throughout which the internal forces cannot be represented by a pressure, will be bounded by a surface very near to the solid, provided the velocity of the solid relative to the liquid at some distance from it is not too great; therefore a sufficiently good representation of the motion will be obtained by taking the above condition to be that which is satisfied at the boundary of the solid, and this representation will be consistent with the fact that no slipping takes place at the surface of the solid, the transition from the solid to the liquid being supposed to take place in a thin layer whose thickness is negligible. When one liquid is in contact with another, there will be a transition layer throughout which the internal forces cannot be represented by a pressure; as in the previous case, provided the velocity of the one liquid relative to the other is not too great, this transition layer can be taken to be very thin, and its effect can be represented by supposing a vortex sheet to coincide with the surface of separation of the liquids. The motions of jets and of heterogeneous liquids belong to this class.

In a liquid the condition, that the same molecules always occupy an element of volume of the same size, determines the nature of the internal constraint, that it can be represented by a pressure. In a fluid this condition does not hold; the relation

$$\rho \, \frac{\partial \, (x, \, y, \, z)}{\partial \, (x_0, \, y_0, \, z_0)} = \rho_0$$

only expresses the fact that the distribution of the molecules is always such that the distance between any two adjacent molecules is very small and that the total number of molecules is unalterable; the nature of the internal forces cannot be determined from this relation. It has to be assumed that the internal forces in a fluid are of the same kind as in a liquid, and that they can be represented by a pressure, the existence of internal forces in the nature of shearing stresses being incompatible with the idea of perfect mobility. This assumption having been made, it is further necessary, in order that the transformation of the function

$$\iiint [\tfrac{1}{2} \rho \, (\dot{x}^2 + \dot{y}^2 + \dot{z}^2) - \rho V + \kappa] \, dx \, dy \, dz$$

from the space occupied by the fluid at time t to the space occupied by it at time t_0 can be performed, that κ should be a function of ρ only. When this further condition is satisfied, the variation can be performed, and the equations of motion are the same as in the case of a liquid, so that the statements made above concerning the motions of liquids apply equally to the motions of fluids. Just as in the case of the elastic solid, the application of continuous analysis to obtain the equations of motion involves the assumption that the changes which take place in the fluid take place adiabatically. Thus, it being assumed that a continuous analysis can be applied to determine the circumstances of the propagation of waves of sound, the law connecting pressure and density must be taken to be the adiabatic law.

When a mass of fluid is in motion, the course of events may be such that the fluid, supposed initially to occupy a

space bounded by one closed surface, tends to divide itself
into two separate portions, each occupying a space bounded
by a closed surface. It is at once clear that a continuous
analysis cannot be applied to trace the history of the changes
that take place while the fluid passes from the state in which
it consists of one portion, to the state in which it consists of
two portions, as during the transition the distances between some
pairs of previously adjacent molecules will become sensible. The
mathematical functions, which express the circumstances of the
motion, will become discontinuous as the instant of separation
is approached, and will thus determine the limits up to which
a continuous analysis can be applied, but it is probable that
the analysis ceases to be applicable some time before this limit
is reached, owing to change in the order of arrangement of the
molecules. A somewhat similar state of affairs exists when
plane waves of condensation and rarefaction, the amplitude of
these waves being finite, are being propagated through an
elastic fluid. The mathematical expressions which, in this
case, specify the course of events can be obtained. If u denote
the velocity and y the position of a molecule of the fluid at
time t, the relation is

$$u = f\{y - (a + u(1 + \gamma)/2)\, t\},$$

where γ is the ratio of the specific heats of the fluid and a is
the velocity of propagation of waves of small amplitude. From
this it appears that, as the wave advances, the condensation in
one part of the wave and the rarefaction in another part of the
wave increase without limit; the above expression will therefore cease to represent what is going on in an actual fluid some
time before the expression becomes discontinuous, either because
the distance between adjacent molecules in the rarefied part of
the fluid has become sensible, or because the law connecting
pressure and density no longer represents the internal changes,
whichever of the two should first happen.

From the above discussion of material media it would
appear that the process known as " treating them as continuous " is one of approximation, the limits of accuracy of the
approximation depending on the distances between adjacent
pairs of molecules; and the comparison of the results of obser-

vation and theory shews that, within limits, the treatment gives sufficiently accurate results. If, instead of being regarded as a method of approximation suitable for the treatment of material media, the above analysis be regarded as representing the changes which take place in a truly continuous medium the assumption is involved that the notion of mass can be transferred from a material medium to a truly continuous medium; but, as associated with a truly continuous medium, this notion involves the actual attainment of the limit ρ for the "density at a point" (cf. p. 146, *supra*) and it further involves the assumption of the existence at each point of a vector u, v, w in terms of which the kinetic energy of the continuous medium can be represented by an expression of the form

$$\iiint \tfrac{1}{2}\rho \, (u^2 + v^2 + w^2) \, dx \, dy \, dz.$$

A continuous medium which is such that its kinetic energy can be represented in this way is not of the most general type of continuous medium conceivable. To distinguish it from other possible continuous media such a medium may be said to possess atomic structure. It has been seen that, when a material medium is in motion subject to the laws of dynamics, the condition that the medium should continue to exist as a medium to which continuous analysis can be applied requires restrictions to be placed on the possible motions. These restrictions have to be provided for by the existence of internal forces in the medium, such internal forces not being forces of motion of the medium. The question then naturally arises, what is the origin of these internal forces? Two answers can be given to this question, that these internal forces are inherent in the medium, or that they are due to the motions of some other medium which occupies the same space. The statement, that the internal forces are inherent in the medium, is equivalent to the statement that their origin is unknown. There remains then the hypothesis that the internal forces of a material medium are due to the motions of some other medium which occupies the same space. In view of the fact that experience of physical changes is confined to the comparison of motions it would seem natural to expect that the energy of the ultimate medium is all kinetic energy, but a continuous medium

which possesses atomic structure cannot, in virtue of its own
motions, produce any effect on the motions of a material medium
occupying the same space, for such media, so far as their motions
are concerned, are dynamically independent systems.

It follows therefore that, if the internal forces of a material
medium are due to the presence of another medium in the
same space, this other medium is not a continuous medium
possessing atomic structure, that is, its kinetic energy is not
expressible in the form $\iiint \frac{1}{2}\rho\,(u^2 + v^2 + w^2)\,dx\,dy\,dz$. It must
therefore be a continuous medium of a different type. That a
continuous medium may exist which is capable of providing
the necessary internal forces in a material medium which is
imbedded in it would appear to follow from geometrical considerations, for a continuous medium can be imagined which is
such that there is no point in it which is not affected by a
change taking place at any other point in it. The laws of
motion of such a medium, if they were known, would be
expressed geometrically, and, in order to submit the circumstances of the motions to calculation, it may be assumed that,
although these laws are not known, the motions possess a
Lagrangian function, this function being a homogeneous
quadratic function of the time rates of variation of all the
coordinates which are necessary for the specification of the
geometrical changes which can take place in any small volume.
It need not be assumed that this Lagrangian function completely represents the motions of the aether; it is sufficient
to assume that it represents the effects of the motions of
the aether so far as they affect the motions of arithmetically
continuous media. This is the assumption that has been
made in Chapter v. as to the Lagrangian function of the
motions of the aether, and it has been shewn there that the
modified Lagrangian function, which is arrived at in the
Faraday-Maxwell theory of the electrical behaviour of the
aether and which involves only those coordinates whose changes
are associated with electrical effects in material media, is consistent with this assumption. This modified Lagrangian function
is the analytical representation of Faraday's laws, and in it
the axes of reference of the motions are those axes for which
Faraday's laws are accurately true.

APPENDIX C.

THE ELECTRODYNAMICS OF MOVING MEDIA.

In Chapter II. the equations of electrodynamics were obtained directly from Faraday's laws, and these equations involved a certain vector F, G, H, which was there obtained in terms of the convection currents. It was remarked that this vector F, G, H differed from Maxwell's electrokinetic momentum by a vector whose components are the differential coefficients of a scalar function. Maxwell's electrokinetic momentum is defined as the vector whose components F, G, H are double the coefficients of the current strength in the expression for the electrical energy. The application of the Lagrangian method will therefore give equations which will determine F, G, H uniquely and the result will put in evidence the difference between the F, G, H which are the components of the electrokinetic momentum and the F, G, H of Chapter II. The Lagrangian function of the motions of the system, which consists of the aether and moving electric charges, is

$$\tfrac{1}{2} \iiint [F(\dot{f}+u) + G(\dot{g}+v) + H(\dot{h}+w) - Xf - Yg - Zh]\, dx\, dy\, dz + L,$$

where F, G, H are the components of the electrokinetic momentum, X, Y, Z are the components of the electric force, f, g, h are the components of the aethereal electric displacement, u, v, w are the components of the convection current at the point x, y, z, and L is the part of the Lagrangian function which is independent of f, g, h, the axes of reference being those for which Faraday's laws are accurately true. In this expression the coordinates which specify degrees of freedom are the coordinates f, g, h belonging to the aether and the coordinates x, y, z which at any instant of time define the positions of the moving charges. In order that it may be possible to apply the methods of continuous analysis, the moving charges must be supposed to be replaced by a distribution of volume-density ρ at the point x, y, z, such that ρ and its differential coefficients are finite and continuous everywhere, and it will appear later how, after the equations, which determine the history of the changes, have been integrated, a transition can be made from the distribution ρ to the isolated charges. The relations between the other quantities, which occur in the Lagrangian function, and the coordinates are

$$(X,\ Y,\ Z) = 4\pi V^2 (f,\ g,\ h) \ \dots\dots\dots\dots(1),$$

$$(\alpha,\ \beta,\ \gamma) = \left(\frac{\partial H}{\partial y} - \frac{\partial G}{\partial z},\ \frac{\partial F}{\partial z} - \frac{\partial H}{\partial x},\ \frac{\partial G}{\partial x} - \frac{\partial F}{\partial y}\right) \ \dots(2),$$

$$4\pi\,(\dot{f}+u,\ \dot{g}+v,\ \dot{h}+w) = \left(\frac{\partial \gamma}{\partial y} - \frac{\partial \beta}{\partial z},\ \frac{\partial \alpha}{\partial z} - \frac{\partial \gamma}{\partial x},\ \frac{\partial \beta}{\partial x} - \frac{\partial \alpha}{\partial y}\right)\dots(3),$$

$$(u,\ v,\ w) = \rho\,(\dot{x},\ \dot{y},\ \dot{z}) \ \dots\dots\dots\dots\dots\dots(4),$$

where α, β, γ are the components of the magnetic force and V is a constant. Further f, g, h are not independent, but are subject to the relation

$$\frac{\partial f}{\partial x} + \frac{\partial g}{\partial y} + \frac{\partial h}{\partial z} = \rho\dots\dots\dots\dots\dots(5).$$

The application of the Lagrangian method gives the equation

$$\delta \int_{t_0}^{t_1}\!\!\iiint \tfrac{1}{2}\,[F\,(\dot{f}+u) + G\,(\dot{g}+v) + H\,(\dot{h}+w)$$
$$- Xf - Yg - Zh]\,dx\,dy\,dz\,dt + \delta L = 0\dots\dots\dots(6),$$

subject to the relation (5), and, introducing the undetermined multiplier ϕ, which must be a function of x, y, z only and independent of the time, this becomes

$$(\delta_1 + \delta_2) \int_{t_0}^{t_1}\!\!\iiint \left[\tfrac{1}{2}\,\{F\,(\dot{f}+u) + G\,(\dot{g}+v) + H\,(\dot{h}+w)\right.$$
$$\left. - Xf - Yg - Zh\} + \phi\left\{\frac{\partial f}{\partial x} + \frac{\partial g}{\partial y} + \frac{\partial h}{\partial z} - \rho\right\} \right] dx\,dy\,dz\,dt$$
$$+ \delta_2 L = 0 \ \dots\dots\dots\dots\dots\dots(6'),$$

where δ_1 refers to variations of the coordinates f, g, h, and δ_2 to variations of the coordinates x, y, z belonging to the moving charges. The variations δ_1 will give the equations belonging to the aether and the variations δ_2 the equations belonging to the motions of the moving charges. Taking these separately, the part which gives the equations to be satisfied in the aether is, remembering the relations (1), (2), (3),

$$\int_{t_0}^{t_1}\iiint\Big[F\delta\dot{f} + G\delta\dot{g} + H\delta\dot{h} - X\delta f - Y\delta g - Z\delta h$$
$$+ \phi\left\{\frac{\partial\delta f}{\partial x} + \frac{\partial\delta g}{\partial y} + \frac{\partial\delta h}{\partial z}\right\}\Big]\,dx\,dy\,dz\,dt = 0,$$

that is, integrating by parts,

$$\iiint\Big|_{t_0}^{t_1} (F\delta f + G\delta g + H\delta h)\,dx\,dy\,dz$$
$$+ \int_{t_0}^{t_1}\iint \phi\,(\delta f\,dy\,dz + \delta g\,dz\,dx + \delta h\,dx\,dy)\,dt$$
$$- \int_{t_0}^{t_1}\iiint\Big[\left(X + \frac{\partial F}{\partial t} + \frac{\partial\phi}{\partial x}\right)\delta f + \left(Y + \frac{\partial G}{\partial t} + \frac{\partial\phi}{\partial y}\right)\delta g$$
$$+ \left(Z + \frac{\partial H}{\partial t} + \frac{\partial\phi}{\partial z}\right)\delta h\Big]\,dx\,dy\,dz\,dt = 0.$$

The expression

$$\iiint\Big|_{t_0}^{t_1} (F\delta f + G\delta g + H\delta h)\,dx\,dy\,dz$$

vanishes identically on account of the conditions under which the variations are performed. The remaining triple integral and the quadruple integral must separately vanish. The condition, that the triple integral

$$\int_{t_0}^{t_1}\iint \phi\,(\delta f\,dy\,dz + \delta g\,dz\,dx + \delta h\,dx\,dy)\,dt$$

should vanish, requires that ϕ should be continuous everywhere and vanish at the infinitely distant boundary. The condition, that the quadruple integral should vanish, requires that the equations

$$\left.\begin{aligned}
X + \frac{\partial F}{\partial t} + \frac{\partial \phi}{\partial x} &= 0, \\
Y + \frac{\partial G}{\partial t} + \frac{\partial \phi}{\partial y} &= 0, \\
Z + \frac{\partial H}{\partial t} + \frac{\partial \phi}{\partial z} &= 0,
\end{aligned}\right\} \quad \ldots\ldots\ldots\ldots\ldots\ldots(7),$$

should be satisfied at every point.

The equations of motion of the moving charges are given by

$$\delta_2 \int_{t_0}^{t_1} \!\!\!\iiint \left[\tfrac{1}{2} \{ F(\dot{f} + u) + G(\dot{g} + v) + H(\dot{h} + w) \} \right.$$
$$\left. + \phi \left\{ \frac{\partial f}{\partial x} + \frac{\partial g}{\partial y} + \frac{\partial h}{\partial z} - \rho \right\} \right] dx\, dy\, dz\, dt + \delta_2 L = 0,$$

where the coordinates which are now varied are the coordinates which define the moving charges. In order to effect the variation, the integrals must be transformed to an integral taken throughout the same space at each instant, and the space may be chosen to be that occupied by the moving charges at the time t_0. Writing

$$\delta \psi = \delta_2 \int_{t_0}^{t_1} \!\!\!\iiint \tfrac{1}{2} \left[F(\dot{f} + u) + G(\dot{g} + v) + H(\dot{h} + w) \right] dx\, dy\, dz\, dt,$$

and effecting the transformation, this becomes

$$\delta \psi = \delta_2 \int_{t_0}^{t_1} \!\!\!\iiint \tfrac{1}{2} \left[F(\dot{f} + u) + G(\dot{g} + v) + H(\dot{h} + w) \right] \frac{\rho_0}{\rho} dx_0\, dy_0\, dz_0\, dt,$$

where x_0, y_0, z_0 at the time t_0 are the coordinates of the point belonging to a moving charge which, at the time t, occupies the position x, y, z, and ρ_0 is the corresponding volume-density, it being remembered that the total charge is unalterable. Now F, G, H are linear in $\dot{f} + u$, $\dot{g} + v$, $\dot{h} + w$, and therefore, performing the variation,

$$\delta \psi = \int_{t_0}^{t_1} \!\!\!\iiint \left[F\delta \dot{x} + G\delta \dot{y} + H\delta \dot{z} + \left(\dot{x} \frac{\partial F}{\partial x} + \dot{y} \frac{\partial G}{\partial x} + \dot{z} \frac{\partial H}{\partial x} \right) \delta x \right.$$
$$\left. + \left(\dot{x} \frac{\partial F}{\partial y} + \dot{y} \frac{\partial G}{\partial y} + \dot{z} \frac{\partial H}{\partial y} \right) \delta y + \left(\dot{x} \frac{\partial F}{\partial z} + \dot{y} \frac{\partial G}{\partial z} + \dot{z} \frac{\partial H}{\partial z} \right) \delta z \right]$$
$$. \rho_0\, dx_0\, dy_0\, dz_0\, dt,$$

where F, G, H must now be regarded as assuming the succession of values they take as the point x, y, z moves through the aether, not the succession of values they take at a fixed point. Integrating by parts with respect to the time, this becomes

$$\delta\psi = \iiint \Big|_{t_0}^{t_1} [F\delta x + G\delta y + H\delta z]\, \rho_0\, dx_0 dy_0 dz_0$$

$$- \int_{t_0}^{t_1} \iiint \Big[\Big(\frac{\delta F}{\partial t} - \dot{x}\frac{\partial F}{\partial x} - \dot{y}\frac{\partial G}{\partial x} - \dot{z}\frac{\partial H}{\partial x} \Big) \delta x$$

$$+ \Big(\frac{\delta G}{\partial t} - \dot{x}\frac{\partial F}{\partial y} - \dot{y}\frac{\partial G}{\partial y} - \dot{z}\frac{\partial H}{\partial y} \Big) \delta y$$

$$+ \Big(\frac{\delta H}{\partial t} - \dot{x}\frac{\partial F}{\partial z} - \dot{y}\frac{\partial G}{\partial z} - \dot{z}\frac{\partial H}{\partial z} \Big) \delta z \Big]\, \rho_0\, dx_0 dy_0 dz_0 dt,$$

where the symbol $\dfrac{\delta}{\partial t}$ is used to indicate that the time rates of variation of F, G, H are to be calculated on the understanding that they assume the succession of values specified above. Now in the equations (7) F, G, H are regarded as assuming a succession of values at a fixed point, and, as it is by means of these equations that F, G, H have to be determined, it is convenient to regard F, G, H as expressed in this way throughout.   On this understanding

$$\frac{\delta F}{\partial t} = \frac{\partial F}{\partial t} + \dot{x}\frac{\partial F}{\partial x} + \dot{y}\frac{\partial F}{\partial y} + \dot{z}\frac{\partial F}{\partial z} *, \text{ etc.}$$

and therefore

$$\frac{\delta F}{\partial t} - \dot{x}\frac{\partial F}{\partial x} - \dot{y}\frac{\partial G}{\partial x} - \dot{z}\frac{\partial H}{\partial x} = \frac{\partial F}{\partial t} - \dot{y}\gamma + \dot{z}\beta, \text{ etc.};$$

hence

$$\delta\psi = \iiint \Big|_{t_0}^{t_1} [F\delta x + G\delta y + H\delta z]\, \rho_0\, dx_0 dy_0 dz_0$$

$$- \int_{t_0}^{t_1} \iiint \Big[\Big(\frac{\partial F}{\partial t} - \dot{y}\gamma + \dot{z}\beta \Big) \delta x + \Big(\frac{\partial G}{\partial t} - \dot{z}\alpha + \dot{x}\gamma \Big) \delta y$$

$$+ \Big(\frac{\partial H}{\partial t} - \dot{x}\beta + \dot{y}\alpha \Big) \delta z \Big]\, \rho_0\, dx_0 dy_0 dz_0 dt.$$

---

* Lagrange, *Mécanique Analytique*, 3rd edition, II. p. 263.

The triple integral vanishes on account of the conditions under which the variations are performed, and therefore

$$\delta\psi = -\int_{t_0}^{t_1}\iiint\left[\left(\frac{\partial F}{\partial t} - \dot{y}\gamma + \dot{z}\beta\right)\delta x + \left(\frac{\partial G}{\partial t} - \dot{z}\alpha + \dot{x}\gamma\right)\delta y\right.$$
$$\left. + \left(\frac{\partial H}{\partial t} - \dot{x}\beta + \dot{y}\alpha\right)\delta z\right]\rho\,dx\,dy\,dz\,dt.$$

Again, writing

$$\delta\chi = \delta_2\int_{t_0}^{t_1}\iiint\phi\left[\frac{\partial f}{\partial x} + \frac{\partial g}{\partial y} + \frac{\partial h}{\partial z} - \rho\right]dx\,dy\,dz\,dt,$$

and transforming to the space occupied by the charges at the time t_0, it becomes

$$\delta\chi = \delta_2\int_{t_0}^{t_1}\iiint\phi\left[\left\{\frac{\partial f}{\partial x} + \frac{\partial g}{\partial y} + \frac{\partial h}{\partial z}\right\}\frac{\partial(x, y, z)}{\partial(x_0, y_0, z_0)} - \rho_0\right]dx_0\,dy_0\,dz_0\,dt;$$

that is

$$\delta\chi = \int_{t_0}^{t_1}\iiint\phi\left[\frac{\partial f}{\partial x} + \frac{\partial g}{\partial y} + \frac{\partial h}{\partial z}\right]\left[\frac{\partial(\delta x, y, z)}{\partial(x_0, y_0, z_0)}\right.$$
$$\left. + \frac{\partial(x, \delta y, z)}{\partial(x_0, y_0, z_0)} + \frac{\partial(x, y, \delta z)}{\partial(x_0, y_0, z_0)}\right]dx_0\,dy_0\,dz_0\,dt,$$

it being remembered that f, g, h are independent of the coordinates defining the positions of the moving charges. Denoting by $\delta\chi_1$, $\delta\chi_2$, $\delta\chi_3$ the parts of this expression which involve δx, δy, δz respectively,

$$\delta\chi_1 = \int_{t_0}^{t_1}\iiint\phi\left[\frac{\partial f}{\partial x} + \frac{\partial g}{\partial y} + \frac{\partial h}{\partial z}\right]\left[\frac{\partial(y, z)}{\partial(y_0, z_0)}\frac{\partial\delta x}{\partial x_0}\right.$$
$$\left. + \frac{\partial(y, z)}{\partial(z_0, x_0)}\frac{\partial\delta x}{\partial y_0} + \frac{\partial(y, z)}{\partial(x_0, y_0)}\frac{\partial\delta x}{\partial z_0}\right]dx_0\,dy_0\,dz_0\,dt,$$

which, integrating by parts and transforming to the space occupied by the moving charges at the time t, becomes

$$\delta\chi_1 = \int_{t_0}^{t_1}\iint\phi\left(\frac{\partial f}{\partial x} + \frac{\partial g}{\partial y} + \frac{\partial h}{\partial z}\right)\delta x\,dy\,dz$$
$$- \int_{t_0}^{t_1}\iiint\frac{\partial\phi}{\partial x}\left(\frac{\partial f}{\partial x} + \frac{\partial g}{\partial y} + \frac{\partial h}{\partial z}\right)\delta x\,dx\,dy\,dz\,dt.$$

Hence

$$\delta\chi = \int_{t_0}^{t_1}\!\!\iint \phi\left[\delta x\,dy\,dz + \delta y\,dz\,dx + \delta z\,dx\,dy\right]\rho\,dt$$

$$-\int_{t_0}^{t_1}\!\!\iiint\left[\frac{\partial\phi}{\partial x}\,\delta x + \frac{\partial\phi}{\partial y}\,\delta y + \frac{\partial\phi}{\partial z}\,\delta z\right]\rho\,dx\,dy\,dz\,dt,$$

where the triple integral vanishes on account of the conditions which have already been found for ϕ, and therefore

$$\delta\chi = -\int_{t_0}^{t_1}\!\!\iiint\left[\frac{\partial\phi}{\partial x}\,\delta x + \frac{\partial\phi}{\partial y}\,\delta y + \frac{\partial\phi}{\partial z}\,\delta z\right]\rho\,dx\,dy\,dz\,dt.$$

Writing $\qquad\qquad L = \iiint L'\,dx\,dy\,dz,$

there results

$$\delta_2 L = -\iiint\left[\left\{\frac{d}{dt}\left(\frac{\partial L'}{\partial \dot x}\right) - \frac{\partial L'}{\partial x}\right\}\delta x + \ldots + \ldots\right]dx\,dy\,dz,$$

and therefore the equations of motion for the moving charges are

$$\left.\begin{aligned}
\frac{d}{dt}\left(\frac{\partial L'}{\partial \dot x}\right) - \frac{\partial L'}{\partial x} + \rho\left(\frac{\partial F}{\partial t} - \dot y\gamma + \dot z\beta + \frac{\partial\phi}{\partial x}\right) &= 0,\\[4pt]
\frac{d}{dt}\left(\frac{\partial L'}{\partial \dot y}\right) - \frac{\partial L'}{\partial y} + \rho\left(\frac{\partial G}{\partial t} - \dot z\alpha + \dot x\gamma + \frac{\partial\phi}{\partial y}\right) &= 0,\\[4pt]
\frac{d}{dt}\left(\frac{\partial L'}{\partial \dot z}\right) - \frac{\partial L'}{\partial z} + \rho\left(\frac{\partial H}{\partial t} - \dot x\beta + \dot y\alpha + \frac{\partial\phi}{\partial z}\right) &= 0,
\end{aligned}\right\} \cdots (8).$$

Writing

$$\left.\begin{aligned}
X' &= -\frac{\partial F}{\partial t} + \dot y\gamma - \dot z\beta - \frac{\partial\phi}{\partial x},\\[4pt]
Y' &= -\frac{\partial G}{\partial t} + \dot z\alpha - \dot x\gamma - \frac{\partial\phi}{\partial y},\\[4pt]
Z' &= -\frac{\partial H}{\partial t} + \dot x\beta - \dot y\alpha - \frac{\partial\phi}{\partial z},
\end{aligned}\right\}\cdots\cdots\cdots\cdots (9),$$

X', Y', Z' are termed the components of the electric force acting on an element of the moving charge, and the equations (8) become

$$\left.\begin{aligned}
\frac{d}{dt}\left(\frac{\partial L'}{\partial \dot x}\right) - \frac{\partial L'}{\partial x} &= \rho X',\\[4pt]
\frac{d}{dt}\left(\frac{\partial L'}{\partial \dot y}\right) - \frac{\partial L'}{\partial y} &= \rho Y',\\[4pt]
\frac{d}{dt}\left(\frac{\partial L'}{\partial \dot z}\right) - \frac{\partial L'}{\partial z} &= \rho Z',
\end{aligned}\right\}\cdots\cdots\cdots\cdots (8').$$

Before proceeding with the discussion of these equations, it is convenient to obtain, by means of the equations (1), (2), (3), (5) and (7), expressions for F, G, H and the related quantities in terms of the distribution of the moving charges. It follows from equations (2) and (3) that

$$4\pi\,(\dot{f}+u)=\frac{\partial}{\partial x}\left(\frac{\partial F}{\partial x}+\frac{\partial G}{\partial y}+\frac{\partial H}{\partial z}\right)-\frac{\partial^2 F}{\partial x^2}-\frac{\partial^2 F}{\partial y^2}-\frac{\partial^2 F}{\partial z^2}\,,$$

with two similar equations, that is, writing

$$J=\frac{\partial F}{\partial x}+\frac{\partial G}{\partial y}+\frac{\partial H}{\partial z}\,,$$

and

$$\nabla^2=\frac{\partial^2}{\partial x^2}+\frac{\partial^2}{\partial y^2}+\frac{\partial^2}{\partial z^2}\,,$$

$$4\pi\,(\dot{f}+u)=\frac{\partial J}{\partial x}-\nabla^2 F,\ \text{etc.}$$

Again, from (1) and (7),

$$4\pi V^2 f=-\frac{\partial F}{\partial t}-\frac{\partial\phi}{\partial x}\,,\ \text{etc.,}$$

that is, remembering that ϕ is independent of the time,

$$4\pi V^2\dot{f}=-\frac{\partial^2 F}{\partial t^2}\,,\ \text{etc.,}$$

and therefore

$$\nabla^2 F-\frac{1}{V^2}\frac{\partial^2 F}{\partial t^2}+4\pi u-\frac{\partial J}{\partial x}=0,$$

$$\nabla^2 G-\frac{1}{V^2}\frac{\partial^2 G}{\partial t^2}+4\pi v-\frac{\partial J}{\partial y}=0,$$

$$\nabla^2 H-\frac{1}{V^2}\frac{\partial^2 H}{\partial t^2}+4\pi w-\frac{\partial J}{\partial z}=0.$$

The solutions* of these equations are

$$F=F'+\frac{\partial\chi}{\partial x}\,,\quad G=G'+\frac{\partial\chi}{\partial y}\,,\quad H=H'+\frac{\partial\chi}{\partial z}\,,$$

* See § 15 above.

where

$$F' = \iiint \frac{u_1}{r} \, dx_1 dy_1 dz_1, \quad G' = \iiint \frac{v_1}{r} \, dx_1 dy_1 dz_1,$$

$$H' = \iiint \frac{w_1}{r} \, dx_1 dy_1 dz_1,$$

r being the distance of the point x, y, z from the point x_1, y_1, z_1, and u_1, v_1, w_1 denoting the values of u, v, w at the point x_1, y_1, z_1 at the time $t - r/V$; further χ satisfies the equation

$$\frac{1}{V^2} \frac{\partial^2 \chi}{\partial t^2} + \frac{\partial F'}{\partial x} + \frac{\partial G'}{\partial y} + \frac{\partial H'}{\partial z} = 0.$$

Now, writing

$$J' = \frac{\partial F'}{\partial x} + \frac{\partial G'}{\partial y} + \frac{\partial H'}{\partial z},$$

$$J' = \iiint \left[\frac{\partial}{\partial x} \frac{u_1}{r} + \frac{\partial}{\partial y} \frac{v_1}{r} + \frac{\partial}{\partial z} \frac{w_1}{r} \right] dx_1 dy_1 dz_1,$$

that is

$$J' = \iiint \left[u_1 \frac{\partial}{\partial x} \frac{1}{r} + \frac{1}{r} \frac{\partial u_1}{\partial x} + \ldots + \ldots \right] dx_1 dy_1 dz_1,$$

hence

$$J' = \iiint \frac{1}{r} \left[\left(\frac{\partial u_1}{\partial x_1} + \frac{\partial u_1}{\partial x} \right) + \left(\frac{\partial v_1}{\partial y_1} + \frac{\partial v_1}{\partial y} \right) + \left(\frac{\partial w_1}{\partial z_1} + \frac{\partial w_1}{\partial z} \right) \right] dx_1 dy_1 dz_1,$$

for $\dfrac{\partial}{\partial x} \dfrac{1}{r} = - \dfrac{\partial}{\partial x_1} \dfrac{1}{r}$ and u, v, w are continuous.

The unalterability of the total charge gives the relation

$$\frac{\partial \rho_1}{\partial t} + \frac{\partial u_1}{\partial x_1} + \frac{\partial u_1}{\partial x} + \frac{\partial v_1}{\partial y_1} + \frac{\partial v_1}{\partial y} + \frac{\partial w_1}{\partial z_1} + \frac{\partial w_1}{\partial z} = 0,$$

where ρ_1 denotes the value of ρ at the point x_1, y_1, z_1 at the time $t - r/V$, therefore

$$J' = - \iiint \frac{1}{r} \frac{\partial \rho_1}{\partial t} \, dx_1 dy_1 dz_1.$$

Hence

$$\frac{\partial^2 \chi}{\partial t^2} = V^2 \iiint \frac{1}{r} \frac{\partial \rho_1}{\partial t} \, dx_1 dy_1 dz_1,$$

and therefore

$$\frac{\partial \chi}{\partial t} = V^2 \iiint \frac{\rho_1'}{r} \, dx_1 dy_1 dz_1,$$

where ρ' denotes the volume-density of that part of the distribution of electric charge which depends on the time, that is the part of the distribution which is moving. It follows from equations (1), (5) and (7) that

$$4\pi V^2 \rho = -\frac{\partial J}{\partial t} - \nabla^2 \phi,$$

that is

$$4\pi V^2 \rho + \left(\nabla^2 - \frac{1}{V^2}\frac{\partial^2}{\partial t^2}\right)\frac{\partial \chi}{\partial t} + \nabla^2 \phi = 0;$$

whence

$$4\pi V^2 (\rho - \rho') + \nabla^2 \phi = 0.$$

Writing $\rho - \rho' = \rho''$, this being the volume-density of the part of the distribution of electric charge which does not move, ϕ is given by

$$\phi = V^2 \iiint \frac{\rho''}{r}\, dx_1 dy_1 dz_1.$$

The components of the electric force are therefore given by

$$
\begin{aligned}
X &= -\iiint \left[\frac{1}{r}\frac{\partial u_1}{\partial t} + V^2\frac{\partial}{\partial x}\left(\frac{\rho_1' + \rho''}{r}\right)\right] dx_1 dy_1 dz_1, \\
Y &= -\iiint \left[\frac{1}{r}\frac{\partial v_1}{\partial t} + V^2\frac{\partial}{\partial y}\left(\frac{\rho_1' + \rho''}{r}\right)\right] dx_1 dy_1 dz_1, \\
Z &= -\iiint \left[\frac{1}{r}\frac{\partial w_1}{\partial t} + V^2\frac{\partial}{\partial z}\left(\frac{\rho_1' + \rho''}{r}\right)\right] dx_1 dy_1 dz_1,
\end{aligned}
\quad \dots (10),
$$

and the components of the magnetic force by

$$
\begin{aligned}
\alpha &= \iiint \left[\frac{\partial}{\partial y}\frac{w_1}{r} - \frac{\partial}{\partial z}\frac{v_1}{r}\right] dx_1 dy_1 dz_1, \\
\beta &= \iiint \left[\frac{\partial}{\partial z}\frac{u_1}{r} - \frac{\partial}{\partial x}\frac{w_1}{r}\right] dx_1 dy_1 dz_1, \\
\gamma &= \iiint \left[\frac{\partial}{\partial x}\frac{v_1}{r} - \frac{\partial}{\partial y}\frac{u_1}{r}\right] dx_1 dy_1 dz_1,
\end{aligned}
\quad \dots \dots \dots (11).*
$$

Proceeding now to the discussion of special cases of moving distributions, the first case to be considered is that in which an electrical distribution is moving with a uniform velocity v in a given direction. Choosing this direction as the axis of x, the

* Cf. Levi-Civita, *Il Nuovo Cimento*, S. 4, t. vi. 1897.

coordinates of any point of the moving distribution at the time t will be x_1, y_1, z_1, where

$$x_1 = x_0 + vt, \quad y_1 = y_0, \quad z_1 = z_0,$$

x_0, y_0, z_0 denoting the coordinates of the same point at the time from which t is measured.

In this case ρ'' is zero and the expressions to be evaluated are

$$\iiint \frac{u_1}{r'} dx_1' dy_1' dz_1', \quad \iiint \frac{v_1}{r'} dx_1' dy_1' dz_1', \quad \iiint \frac{w_1}{r'} dx_1' dy_1' dz_1',$$

$$\iiint \frac{\rho_1}{r'} dx_1' dy_1' dz_1',$$

where

$$r'^2 = (x - x_1')^2 + (y - y_1')^2 + (z - z_1')^2,$$

and u_1, v_1, w_1, ρ_1 denote the values of u, v, w, ρ at the point x_1', y_1', z_1' at the time $t - r'/V$, this being the point of the distribution which was at the point x_0, y_0, z_0 initially. Then

$$x_1' = x_0 + v\left(t - \frac{r'}{V}\right), \quad y_1' = y_0, \quad z_1' = z_0,$$

that is

$$x_1' = x_1 - \frac{vr'}{V}, \quad y_1' = y_1, \quad z_1' = z_1.$$

Now $v = 0$, $w = 0$, therefore

$$\iiint \frac{v_1}{r'} dx_1' dy_1' dz_1' = 0, \quad \iiint \frac{w_1}{r'} dx_1' dy_1' dz_1' = 0,$$

and the remaining two integrals have now to be transformed so that they are expressed in terms of the positions of the distribution at the time t. Writing $\kappa = v/V$, r' is given by

$$r'^2 = (x - x_1 + \kappa r')^2 + (y - y_1)^2 + (z - z_1)^2,$$

whence

$$r' = \frac{\kappa}{1 - \kappa^2}(x - x_1) + \frac{R}{\sqrt{(1 - \kappa^2)}},$$

where

$$R^2 = \frac{(x - x_1)^2}{1 - \kappa^2} + (y - y_1)^2 + (z - z_1)^2.$$

The element of volume $dx_1'dy_1'dz_1'$ is that which was occupied at times $t - r'/V$ by the distribution which now occupies the element $dx_1dy_1dz_1$, these times being different for different points of the distribution ; hence

$$u_1 dx_1'dy_1'dz_1' = \rho v \frac{\partial (x_1', y_1', z_1')}{\partial (x_1, y_1, z_1)} dx_1 dy_1 dz_1.$$

Now
$$\frac{\partial x_1'}{\partial x_1} = 1 - \kappa \frac{\partial r'}{\partial x_1},$$

$$r' \frac{\partial r'}{\partial x_1} = - (x - x_1') \frac{\partial x_1'}{\partial x_1},$$

hence
$$\frac{\partial x_1'}{\partial x_1} \left[1 - \frac{\kappa}{r'} (x - x_1') \right] = 1,$$

that is
$$\frac{\partial x_1'}{\partial x_1} \left[1 - \kappa^2 - \frac{\kappa}{r'} (x - x_1) \right] = 1,$$

therefore
$$\frac{\partial x_1'}{\partial x_1} = \frac{r'}{R \sqrt{(1 - \kappa^2)}}.$$

Further,
$$\frac{\partial x_1'}{\partial y_1} = \kappa \frac{y - y_1}{r'}, \qquad \frac{\partial x_1'}{\partial z_1} = \kappa \frac{z - z_1}{r'},$$

$$\frac{\partial y_1'}{\partial x_1} = 0, \qquad \frac{\partial y_1'}{\partial y_1} = 1, \qquad \frac{\partial z_1'}{\partial y_1} = 0, \qquad \frac{\partial z_1'}{\partial z_1} = 1 ;$$

therefore
$$\frac{\partial (x_1', y_1', z_1')}{\partial (x_1, y_1, z_1)} = \frac{r'}{R \sqrt{(1 - \kappa^2)}}.$$

Hence

$$\iiint \frac{u_1}{r'} dx_1'dy_1'dz_1' = \iiint \frac{\rho v}{R \sqrt{(1 - \kappa^2)}} dx_1 dy_1 dz_1,$$

$$\iiint \frac{\rho_1}{r'} dx_1'dy_1'dz_1' = \iiint \frac{\rho}{R \sqrt{(1 - \kappa^2)}} dx_1 dy_1 dz_1.$$

Writing
$$\psi = \frac{1}{\sqrt{(1 - \kappa^2)}} \iiint \frac{\rho}{R} dx_1 dy_1 dz_1,$$

and substituting in equations (10) and (11), the components of the electric and magnetic forces are given by

$$X = - v \frac{\partial \psi}{\partial t} - V^2 \frac{\partial \psi}{\partial x}, \quad Y = - V^2 \frac{\partial \psi}{\partial y}, \quad Z = - V^2 \frac{\partial \psi}{\partial z},$$

that is, remembering that

$$\frac{\partial}{\partial t}\frac{1}{R} = v\frac{\partial}{\partial x_1}\frac{1}{R} = -v\frac{\partial}{\partial x}\frac{1}{R},$$

$$X = -V^2(1 - \kappa^2)\frac{\partial \psi}{\partial x}, \quad Y = -V^2\frac{\partial \psi}{\partial y}, \quad Z = -V^2\frac{\partial \psi}{\partial z} \ldots(12),$$

$$\alpha = 0, \quad \beta = v\frac{\partial \psi}{\partial z}, \quad \gamma = -v\frac{\partial \psi}{\partial y} \ldots\ldots\ldots(13),$$

where ψ satisfies the differential equation

$$(1 - \kappa^2)\frac{\partial^2 \psi}{\partial x^2} + \frac{\partial^2 \psi}{\partial y^2} + \frac{\partial^2 \psi}{\partial z^2} + 4\pi\rho = 0 \ldots\ldots(14)^*.$$

The case of a charged conductor moving through the aether with a uniform velocity can be deduced from the above. The electrical distribution must now be supposed to be the limit of a distribution throughout a thin layer over the surface of the conductor, this distribution being such that ρ and its first differential coefficients tend to zero at the boundary of this layer. The components of the electric and magnetic forces at any point external to the conductor will then be given by equations (12) and (13), and will vanish at an internal point, where ψ now satisfies the differential equation

$$(1 - \kappa^2)\frac{\partial^2 \psi}{\partial x^2} + \frac{\partial^2 \psi}{\partial y^2} + \frac{\partial^2 \psi}{\partial z^2} = 0$$

throughout the external space and is constant over the bounding surface of the conductor.

A simple example is that of a charged conducting sphere of radius a moving with uniform velocity v through the aether. In this case ψ is given by

$$\psi = \frac{E}{\kappa a} \log \coth \frac{\eta}{2},$$

where

$$x^2 \operatorname{sech}^2 \eta + (1 - \kappa^2)(y^2 + z^2)\operatorname{cosech}^2 \eta = \kappa^2 a^2,$$

and E is the charge on the sphere. The surface density of the distribution over the surface of the sphere is uniform.

---

* Lorentz, *Versuch einer theorie der electrischen und optischen erscheinungen in bewegten körpern*, § 22, Leiden 1895. Larmor, *Phil. Trans.* A, 1897; *Aether and Matter*, ch. IX., 1900.

The next case to be considered is that in which the electrical distribution moves in such a way that any point of it describes an orbit which is periodic with reference to some fixed point and whose dimensions are very small. The coordinates of any point of the distribution at the time t will be $x_1 + \xi$, $y_1 + \eta$, $z_1 + \zeta$, where x_1, y_1, z_1 are the coordinates of the point about which it describes a small periodic orbit, and the expressions to be evaluated will be

$$\iiint \frac{u_1}{r_1} d\xi_1 d\eta_1 d\zeta_1, \quad \iiint \frac{v_1}{r_1} d\xi_1 d\eta_1 d\zeta_1,$$

$$\iiint \frac{w_1}{r_1} d\xi_1 d\eta_1 d\zeta_1, \quad \iiint \frac{\rho_1}{r_1} d\xi_1 d\eta_1 d\zeta_1,$$

where

$$r_1{}^2 = (x - x_1 - \xi_1)^2 + (y - y_1 - \eta_1)^2 + (z - z_1 - \zeta_1)^2,$$

and $\xi_1, \eta_1, \zeta_1, u_1, v_1, w_1, \rho_1$ are the values of $\xi, \eta, \zeta, u, v, w, \rho$ at the time $t - r_1/V$. Now, if $\xi = f(t)$, then $\xi_1 = f(t - r_1/V)$, that is

$$\xi_1 = f\left(t - \frac{r}{V} + \frac{r - r_1}{V}\right),$$

where

$$r^2 = (x - x_1 - \xi)^2 + (y - y_1 - \eta)^2 + (z - z_1 - \zeta)^2,$$

and writing

$$\xi' = f\left(t - \frac{r}{V}\right), \text{ etc.,}$$

$$r'^2 = (x - x_1 - \xi')^2 + (y - y_1 - \eta')^2 + (z - z_1 - \zeta')^2,$$

$$r - r' = V\tau,$$

it follows that

$$\xi_1 = \xi' + \dot{\xi}'\tau, \quad \eta_1 = \eta' + \dot{\eta}'\tau, \quad \zeta_1 = \zeta' + \dot{\zeta}'\tau,$$

it being assumed that $(\bar{\xi}/\lambda)^2$, etc. are negligible where $\bar{\xi}$, etc. denote the maximum amplitudes corresponding to a period for which the wave length in the aether is λ. To the same order of magnitude

$$\frac{\partial (\xi_1, \eta_1, \zeta_1)}{\partial (\xi, \eta, \zeta)} = 1 - \frac{\dot{\xi}_1}{V}\frac{\partial r_1}{\partial \xi_1} - \frac{\dot{\eta}_1}{V}\frac{\partial r_1}{\partial \eta_1} - \frac{\dot{\zeta}_1}{V}\frac{\partial r_1}{\partial \zeta_1},$$

that is

$$\frac{\partial\,(\xi_1,\,\eta_1,\,\zeta_1)}{\partial\,(\xi,\,\eta,\,\zeta)} = 1 - \frac{\partial\xi_1}{\partial x} - \frac{\partial\eta_1}{\partial y} - \frac{\partial\zeta_1}{\partial z},$$

and therefore, to this order of magnitude,

$$\iiint \frac{u_1}{r_1}\,d\xi_1 d\eta_1 d\zeta_1 = \iiint \frac{\rho_1\dot{\xi}_1}{r_1}\left(1 - \frac{\partial\xi_1}{\partial x} - \frac{\partial\eta_1}{\partial y} - \frac{\partial\zeta_1}{\partial z}\right) d\xi d\eta d\zeta,$$

$$\iiint \frac{v_1}{r_1}\,d\xi_1 d\eta_1 d\zeta_1 = \iiint \frac{\rho_1\dot{\eta}_1}{r_1}\left(1 - \frac{\partial\xi_1}{\partial x} - \frac{\partial\eta_1}{\partial y} - \frac{\partial\zeta_1}{\partial z}\right) d\xi d\eta d\zeta,$$

$$\iiint \frac{w_1}{r_1}\,d\xi_1 d\eta_1 d\zeta_1 = \iiint \frac{\rho_1\dot{\zeta}_1}{r_1}\left(1 - \frac{\partial\xi_1}{\partial x} - \frac{\partial\eta_1}{\partial y} - \frac{\partial\zeta_1}{\partial z}\right) d\xi d\eta d\zeta,$$

$$\iiint \frac{\rho_1}{r_1}\,d\xi_1 d\eta_1 d\zeta_1 = \iiint \frac{\rho_1}{r_1}\left(1 - \frac{\partial\xi_1}{\partial x} - \frac{\partial\eta_1}{\partial y} - \frac{\partial\zeta_1}{\partial z}\right) d\xi d\eta d\zeta,$$

where, on the right-hand side, ξ_1, η_1, ζ_1, $\dot{\xi}_1$, $\dot{\eta}_1$, $\dot{\zeta}_1$, r_1, ρ_1 now denote the values of ξ, η, ζ, $\dot{\xi}$, $\dot{\eta}$, $\dot{\zeta}$, r, ρ corresponding to the time $t - \dfrac{r'}{V}$. The expressions for the components of the electric and magnetic forces are then obtained by substituting the above expressions in the equations (10) and (11).

The electrical distribution in the above investigation has been assumed to be such that the equations (10) and (11) are applicable. Now the validity of the processes involved in the investigation, by means of which these equations were arrived at, is secured, provided the distribution is such that ρ and its first differential coefficients are everywhere continuous. The transition from a volume distribution to a distribution of isolated electric charges can then be effected in the following way. Let P be a point at which there is an electric charge e, and let a closed surface be drawn enclosing P and no other point at which there is an electric charge, which is always possible as the charges are isolated. Then throughout the volume enclosed by this surface there may be supposed to be a distribution of volume-density ρ, which is such that ρ and its first differential coefficients are continuous, tend to zero as the bounding surface is approached, and vanish at every point on it, and which is such that $\iiint \rho\,dx\,dy\,dz = e$. The distribution arrived at, when this has been done for each one of the charges, is such that the preceding investigations are applicable to it, and the results obtained will hold however small the volumes enclosed by these surfaces may be chosen to be, provided that throughout them ρ can be chosen so as to satisfy the conditions specified above. The electric and magnetic forces due to a distribution of isolated electric charges may therefore be taken to be those given by equations (10) and (11), when the integrals on the right-hand side are replaced by the sums of the limits of those integrals taken throughout the small volumes, these volumes being supposed to be ultimately evanescent, the limits being proceeded to on the supposition, that ρ and its first differential coefficients are continuous throughout any one of these small volumes, vanish at their boundaries, and that $\iiint \rho \, dx \, dy \, dz = e$.

The electric and magnetic forces due to a number of electric charges, each of which is describing, with respect to some fixed point which is not necessarily the same for different charges, a periodic orbit of small linear dimensions, can be obtained from the preceding results by evaluating the limit of the integrals

$$\iiint \frac{\rho_1 \xi_1}{r_1} \left(1 - \frac{\partial \xi_1}{\partial x} - \frac{\partial \eta_1}{\partial y} - \frac{\partial \zeta_1}{\partial z} \right) d\xi \, d\eta \, d\zeta, \text{ etc.}$$

Now it appears from § 48 that the amount of the energy radiated from the distribution depends only on those parts of electric and magnetic forces which, at a great distance, are of the order $1/r$; hence, no part which can contribute to the radiation being omitted, the previous integrals are given by

$$\iiint \frac{u_1}{r_1} d\xi_1 d\eta_1 d\zeta_1 = \iiint \frac{\rho}{r} \frac{\partial \xi_1}{\partial t} d\xi \, d\eta \, d\zeta,$$

$$\iiint \frac{v_1}{r_1} d\xi_1 d\eta_1 d\zeta_1 = \iiint \frac{\rho}{r} \frac{\partial \eta_1}{\partial t} d\xi \, d\eta \, d\zeta,$$

$$\iiint \frac{w_1}{r_1} d\xi_1 d\eta_1 d\zeta_1 = \iiint \frac{\rho}{r} \frac{\partial \zeta_1}{\partial t} d\xi \, d\eta \, d\zeta,$$

$$\iiint \frac{\rho_1}{r_1} d\xi_1 d\eta_1 d\zeta_1 = \iiint \left[\frac{\rho}{r} - \frac{\partial}{\partial x} \frac{\rho \xi_1}{r} - \frac{\partial}{\partial y} \frac{\rho \eta_1}{r} - \frac{\partial}{\partial z} \frac{\rho \zeta_1}{r} \right] d\xi \, d\eta \, d\zeta.$$

Therefore, proceeding to the limit, the values of the electric and magnetic forces, so far as radiation from the distribution is concerned, may be written

$$X = V^2 \left\{ \frac{\partial^2}{\partial x^2} \Sigma \frac{e\xi_1}{r} + \frac{\partial^2}{\partial x \partial y} \Sigma \frac{e\eta_1}{r} + \frac{\partial^2}{\partial x \partial z} \Sigma \frac{e\zeta_1}{r} \right\} - \frac{\partial^2}{\partial t^2} \Sigma \frac{e\xi_1}{r},$$

$$Y = V^2 \left\{ \frac{\partial^2}{\partial y \partial x} \Sigma \frac{e\xi_1}{r} + \frac{\partial^2}{\partial y^2} \Sigma \frac{e\eta_1}{r} + \frac{\partial^2}{\partial y \partial z} \Sigma \frac{e\zeta_1}{r} \right\} - \frac{\partial^2}{\partial t^2} \Sigma \frac{e\eta_1}{r},$$

$$Z = V^2 \left\{ \frac{\partial^2}{\partial z \partial x} \Sigma \frac{e\xi_1}{r} + \frac{\partial^2}{\partial z \partial y} \Sigma \frac{e\eta_1}{r} + \frac{\partial^2}{\partial z^2} \Sigma \frac{e\zeta_1}{r} \right\} - \frac{\partial^2}{\partial t^2} \Sigma \frac{e\zeta_1}{r},$$

$$\alpha = \frac{\partial^2}{\partial y \partial t} \Sigma \frac{e\zeta_1}{r} - \frac{\partial^2}{\partial z \partial t} \Sigma \frac{e\eta_1}{r},$$

$$\beta = \frac{\partial^2}{\partial z \partial t} \Sigma \frac{e\xi_1}{r} - \frac{\partial^2}{\partial x \partial t} \Sigma \frac{e\zeta_1}{r},$$

$$\gamma = \frac{\partial^2}{\partial x \partial t} \Sigma \frac{e\eta_1}{r} - \frac{\partial^2}{\partial y \partial t} \Sigma \frac{e\xi_1}{r},$$

where ξ_1, η_1, ζ_1 now denote the values of ξ, η, ζ at the time $t - r/V$, and

$$r^2 = (x - x_1)^2 + (y - y_1)^2 + (z - z_1)^2.$$

Thus a number of electric charges, moving in the manner specified above, are, in respect of the radiation from them, equivalent to a number of small Hertzian oscillators. Further, if it be assumed that the dimensions of the orbit are so small that $\bar{\xi}^2$ etc. can be neglected in comparison with the square of any length considered, an assumption which is reasonable so far as the applications in view are concerned, the values of the electric and magnetic forces due to the distribution of moving charges are given by

$$X = -\frac{\partial^2}{\partial t^2} \Sigma \frac{e\xi_1}{r} - V^2 \frac{\partial \psi}{\partial x}, \quad Y = -\frac{\partial^2}{\partial t^2} \Sigma \frac{e\eta_1}{r} - V^2 \frac{\partial \psi}{\partial y},$$

$$Z = -\frac{\partial^2}{\partial t^2} \Sigma \frac{e\zeta_1}{r} - V^2 \frac{\partial \psi}{\partial z} \quad \ldots\ldots(15),$$

$$\alpha = \frac{\partial^2}{\partial y \partial t} \Sigma \frac{e\zeta_1}{r} - \frac{\partial^2}{\partial z \partial t} \Sigma \frac{e\eta_1}{r}, \quad \beta = \frac{\partial^2}{\partial z \partial t} \Sigma \frac{e\xi_1}{r} - \frac{\partial^2}{\partial x \partial t} \Sigma \frac{e\zeta_1}{r},$$

$$\gamma = \frac{\partial^2}{\partial x \partial t} \Sigma \frac{e\eta_1}{r} - \frac{\partial^2}{\partial y \partial t} \Sigma \frac{e\xi_1}{r} \quad \ldots\ldots(16),$$

where $r^2 = (x - x_1)^2 + (y - y_1)^2 + (z - z_1)^2$ and ξ_1, η_1, ζ_1 denote the values of ξ, η, ζ at the time $t - \dfrac{r}{V}$.

Taking now the case where there are a number of electric charges, each of which is describing, with respect to some point O, a periodic orbit of small linear dimensions, and where the points O are all moving with the same uniform velocity v in a given direction, which may be taken to be that of the axis of x, this distribution may be replaced by a volume distribution of the kind specified above. Denoting by x_1, y_1, z_1 the co-ordinates of one of the points O at the time t, by x_1', y_1, z_1 the coordinates of the same point at the time $t - r_1/V$, by $x_1 + \xi$, $y_1 + \eta$, $z_1 + \zeta$ the coordinates at the time t of a point of the distribution which is describing, with respect to the point x_1, y_1, z_1, a small periodic orbit, and by $x_1' + \xi_1$, $y_1 + \eta_1$, $z_1 + \zeta_1$ the coordinates of the same point at the time $t - r_1/V$, where

$$r_1^2 = (x - x_1' - \xi_1)^2 + (y - y_1 - \eta_1)^2 + (z - z_1 - \zeta_1)^2,$$

x, y, z being the coordinates of any other point, the integrals on the right-hand side of the equations (10) and (11) which, in this case, have to be evaluated are

$$\iiint \frac{\rho_1'(\dot{x}_1' + \dot{\xi}_1')}{r_1}\, d\xi_1'd\eta_1'd\zeta_1', \qquad \iiint \frac{\rho_1'\dot{\eta}_1'}{r_1}\, d\xi_1'd\eta_1'd\zeta_1',$$

$$\iiint \frac{\rho_1'\dot{\zeta}_1'}{r_1}\, d\xi_1'd\eta_1'd\zeta_1', \qquad \iiint \frac{\rho_1'}{r_1}\, d\xi_1'd\eta_1'd\zeta_1',$$

in which ρ_1', $\dot{\xi}_1'$, $\dot{\eta}_1'$, $\dot{\zeta}_1'$ denote the values of ρ, $\dot{\xi}$, $\dot{\eta}$, $\dot{\zeta}$ at the point $x_1' + \xi_1$, $y_1 + \eta_1$, $z_1 + \zeta_1$ at the time $t - r_1/V$, $d\xi_1'd\eta_1'd\zeta_1'$ denotes the element of volume at that point and $\dot{x}_1' = $ v. Transforming these integrals so that the element of volume $d\xi_1'd\eta_1'd\zeta_1'$ at the point $x_1' + \xi_1$, $y_1 + \eta_1$, $z_1 + \zeta_1$ is replaced by the element of volume $d\xi_1 d\eta_1 d\zeta_1$ at the point $x_1 + \xi_1$, $y_1 + \eta_1$, $z_1 + \zeta_1$*, the Jacobian of the transformation is by a previous investigation given by $r_1/R_1 \sqrt{(1 - \kappa^2)}$, in which $\kappa = $ v$/V$ and

$$R_1^2 = (x - x_1 - \xi_1)^2/(1 - \kappa^2) + (y - y_1 - \eta_1)^2 + (z - z_1 - \zeta_1)^2,$$

---

* The transformation is equivalent to substituting for the fixed axes of reference a set of axes fixed relatively to the moving points O.

and the above integrals become

$$\iiint \frac{\rho_1' \, (\mathrm{v} + \dot{\xi}_1')}{R_1\sqrt{(1-\kappa^2)}} \, d\xi_1 d\eta_1 d\zeta_1, \qquad \iiint \frac{\rho_1' \, \dot{\eta}_1'}{R_1\sqrt{(1-\kappa^2)}} \, d\xi_1 d\eta_1 d\zeta_1,$$

$$\iiint \frac{\rho_1' \, \dot{\zeta}_1'}{R_1\sqrt{(1-\kappa^2)}} \, d\xi_1 d\eta_1 d\zeta_1, \qquad \iiint \frac{\rho_1'}{R_1\sqrt{(1-\kappa^2)}} \, d\xi_1 d\eta_1 d\zeta_1.$$

Now

$$r_1 = \frac{R_1}{\sqrt{(1-\kappa^2)}} + \frac{\kappa}{1-\kappa^2}(x - x_1 - \xi_1),$$

hence, $f(t)$ denoting some function of t,

$$f\left(t - \frac{r_1}{V}\right) = f\left\{t - \frac{R_1}{V\sqrt{(1-\kappa^2)}} - \frac{\kappa(x - x_1 - \xi_1)}{V(1-\kappa^2)}\right\},$$

that is, neglecting κ^3 and higher powers of κ,

$$f\left(t - \frac{r_1}{V}\right) = f\left(t - \frac{R_1}{V}\right) - \frac{\kappa}{V}(x - x_1 - \xi_1)f'\left(t - \frac{R_1}{V}\right)$$
$$+ \frac{\kappa^2}{2V^2}(x - x_1 - \xi_1)^2 f''\left(t - \frac{R_1}{V}\right) - \frac{\kappa^2}{2V}R_1 f'\left(t - \frac{R_1}{V}\right),$$

which may be written

$$f\left(t - \frac{R_1}{V}\right) = f\left(t - \frac{R_1}{V}\right) - \kappa R_1 \frac{\partial f}{\partial \xi_1} + \frac{\kappa^2 R_1}{2V}\frac{\partial}{\partial \xi_1}\{(x - x_1 - \xi_1)f'\}.$$

Then $\dot{\xi}_1$, $\dot{\eta}_1$, $\dot{\zeta}_1$ denoting the values of $\dot{\xi}$, $\dot{\eta}$, $\dot{\zeta}$ at the time $t - R_1/V$, the integral

$$\iiint \frac{\rho_1' \dot{\xi}_1'}{R_1\sqrt{(1-\kappa^2)}} \, d\xi_1 d\eta_1 d\zeta_1 = \iiint \frac{\rho_1 \dot{\xi}_1}{R_1\sqrt{(1-\kappa^2)}} \, d\xi_1 d\eta_1 d\zeta_1$$
$$- \kappa \iiint \frac{\partial}{\partial \xi_1}(\rho_1 \dot{\xi}_1) d\xi_1 d\eta_1 d\zeta_1 + \frac{\kappa^2}{2V}\iiint \frac{\partial}{\partial \xi_1}\{(x - x_1 - \xi_1)(\rho_1 \dot{\xi}_1)'\} d\xi_1 d\eta_1 d\zeta_1,$$

and, since ρ and its first differential coefficients vanish at the boundaries of the small volumes occupied by the distribution which replaces the charges, the second and third integrals on the right-hand side vanish, and therefore

$$\iiint \frac{\rho_1' \dot{\xi}_1'}{R_1\sqrt{(1-\kappa^2)}} \, d\xi_1 d\eta_1 d\zeta_1 = \iiint \frac{\rho_1 \dot{\xi}_1}{R_1\sqrt{(1.-\kappa^2)}} \, d\xi_1 d\eta_1 d\zeta_1,$$

$$\iiint \frac{\rho_1' \dot{\eta}_1'}{R_1\sqrt{(1-\kappa^2)}} \, d\xi_1 d\eta_1 d\zeta_1 = \iiint \frac{\rho_1 \dot{\eta}_1}{R_1\sqrt{(1-\kappa^2)}} \, d\xi_1 d\eta_1 d\zeta_1,$$

$$\iiint \frac{\rho_1' \dot{\zeta}_1'}{R_1\sqrt{(1-\kappa^2)}} \, d\xi_1 d\eta_1 d\zeta_1 = \iiint \frac{\rho_1 \dot{\zeta}_1}{R_1\sqrt{(1-\kappa^2)}} \, d\xi_1 d\eta_1 d\zeta_1.$$

If, as in the previous case, it be assumed that, l denoting the linear dimensions of an orbit, l^2 is negligible in comparison with the square of any length considered, the integrals involved become

$$\frac{1}{\sqrt{(1-\kappa^2)}}\iiint\frac{\rho\dot{\xi_1}}{R}\,d\xi\,d\eta\,d\zeta, \qquad \frac{1}{\sqrt{(1-\kappa^2)}}\iiint\frac{\rho\dot{\eta_1}}{R}\,d\xi\,d\eta\,d\zeta,$$

$$\frac{1}{\sqrt{(1-\kappa^2)}}\iiint\frac{\rho\dot{\zeta_1}}{R}\,d\xi\,d\eta\,d\zeta,$$

$$\frac{1}{\sqrt{(1-\kappa^2)}}\iiint\left[\frac{\rho}{R}-\frac{\partial}{\partial x}\frac{\rho\xi_1}{R}-\frac{\partial}{\partial y}\frac{\rho\eta_1}{R}-\frac{\partial}{\partial z}\frac{\rho\zeta_1}{R}\right]d\xi\,d\eta\,d\zeta,$$

where now

$$R^2 = (x-x_1)^2/(1-\kappa^2)+(y-y_1)^2+(z-z_1)^2,$$

and $\dot{\xi_1}$, $\dot{\eta_1}$, $\dot{\zeta_1}$, ξ_1, η_1, ζ_1 denote the values of $\dot{\xi}$, $\dot{\eta}$, $\dot{\zeta}$, ξ, η, ζ, when $t-R/V$ is written for t. Then writing

$$\psi_1 = \frac{1}{\sqrt{(1-\kappa^2)}}\iiint\frac{\rho\xi_1}{R}\,d\xi\,d\eta\,d\zeta, \qquad \psi_2 = \frac{1}{\sqrt{(1-\kappa^2)}}\iiint\frac{\rho\eta_1}{R}\,d\xi\,d\eta\,d\zeta,$$

$$\psi_3 = \frac{1}{\sqrt{(1-\kappa^2)}}\iiint\frac{\rho\zeta_1}{R}\,d\xi\,d\eta\,d\zeta, \qquad \phi = \frac{1}{\sqrt{(1-\kappa^2)}}\iiint\frac{\rho}{R}\,d\xi\,d\eta\,d\zeta,$$

$$\chi = \frac{\partial\psi_1}{\partial x}+\frac{\partial\psi_2}{\partial y}+\frac{\partial\psi_3}{\partial z},$$

the component of the electric force in the direction of the axis of x is given by

$$X = -\frac{\partial}{\partial t}\frac{1}{\sqrt{(1-\kappa^2)}}\iiint\frac{\rho\dot{\xi_1}}{R}\,d\xi\,d\eta\,d\zeta - \mathrm{v}\frac{\partial}{\partial t}(\phi-\chi)-\mathrm{v}^2\frac{\partial}{\partial x}(\phi-\chi),$$

which is equivalent to

$$X = -\frac{\partial^2\psi_1}{\partial t^2}+\mathrm{v}\frac{\partial^2\psi_1}{\partial x\partial t}+\mathrm{v}\frac{\partial\chi}{\partial t}-(V^2-\mathrm{v}^2)\frac{\partial}{\partial x}(\phi-\chi),$$

in which $\frac{\partial}{\partial t}$ is supposed to operate on ξ_1, η_1, ζ_1 only but not now on x_1. Similarly evaluating the other components, the components of the electric and magnetic forces are given by

$$\left.\begin{aligned}
X &= -\frac{\partial^2\psi_1}{\partial t^2}+\mathrm{v}\frac{\partial^2\psi_1}{\partial x\partial t}+\mathrm{v}\frac{\partial\chi}{\partial t}-V^2(1-\kappa^2)\frac{\partial}{\partial x}(\phi-\chi),\\[4pt]
Y &= -\frac{\partial^2\psi_2}{\partial t^2}+\mathrm{v}\frac{\partial^2\psi_2}{\partial x\partial t}-V^2\frac{\partial}{\partial y}(\phi-\chi),\\[4pt]
Z &= -\frac{\partial^2\psi_3}{\partial t^2}+\mathrm{v}\frac{\partial^2\psi_3}{\partial x\partial t}-V^2\frac{\partial}{\partial z}(\phi-\chi),
\end{aligned}\right\}\dots(17),$$

$$\alpha = \frac{\partial^2 \psi_3}{\partial y \partial t} - \frac{\partial^2 \psi_2}{\partial z \partial t},$$

$$\beta = \frac{\partial^2 \psi_1}{\partial z \partial t} - \frac{\partial^2 \psi_3}{\partial x \partial t} + v \frac{\partial}{\partial z}(\phi - \chi), \Bigg\} \quad \ldots\ldots\ldots\ldots(18).$$

$$\gamma = \frac{\partial^2 \psi_2}{\partial x \partial t} - \frac{\partial^2 \psi_1}{\partial y \partial t} - v \frac{\partial}{\partial y}(\phi - \chi),$$

Returning to equations (8′), integrating them throughout the small volume occupied by the distribution which is supposed to replace a moving charge, and proceeding to the limit, the equations of motion of the moving charge are

$$\frac{d}{dt}\left(\frac{\partial L}{\partial \dot{x}}\right) - \frac{\partial L}{\partial x} = eX',$$

$$\frac{d}{dt}\left(\frac{\partial L}{\partial \dot{y}}\right) - \frac{\partial L}{\partial y} = eY',$$

$$\frac{d}{dt}\left(\frac{\partial L}{\partial \dot{z}}\right) - \frac{\partial L}{\partial z} = eZ',$$

where L is that part of the Lagrangian function which involves only the coordinates that specify the degrees of freedom of the moving charges. The part of L which involves the time rates of variation of these coordinates will be of the form $\Sigma \frac{1}{2} m (\dot{x}^2 + \dot{y}^2 + \dot{z}^2)$, where m is a mass coefficient, these motions being capable of being represented by moving points; the remaining part of L represents the effects other than electrical of the aether on the moving charges, and the equations of motion are therefore

$$m\ddot{x} = X_1 + eX', \Bigg\}$$
$$m\ddot{y} = Y_1 + eY', \Bigg\} \quad \ldots\ldots\ldots\ldots\ldots(19),$$
$$m\ddot{z} = Z_1 + eZ', \Bigg\}$$

where X_1, Y_1, Z_1 represent the non-electrical forces and X', Y', Z' are given by equations (9).

When the charges are supposed to be arranged in permanent groups, each group being such that there is no radiation of energy away from it, the method of Appendix B* can

* p. 149.

be applied to obtain the equations of motion of a group. From the result obtained there it follows that, if x, y, z are now coordinates which specify the position of a group, the equations of motion of the group are

$$\begin{aligned} M\ddot{x} &= \mathbf{X}_1 + \mathbf{X}, \\ M\ddot{y} &= \mathbf{Y}_1 + \mathbf{Y}, \\ M\ddot{z} &= \mathbf{Z}_1 + \mathbf{Z}, \end{aligned} \right\} \quad \dots \dots \dots \dots \dots (20),$$

where $M = \Sigma m$, $\mathbf{X}_1$, $\mathbf{Y}_1$, $\mathbf{Z}_1$ represent the non-electrical forces and $\mathbf{X}$, $\mathbf{Y}$, $\mathbf{Z}$ the electrical forces acting on the group, and $\mathbf{X}_1$, $\mathbf{Y}_1$, $\mathbf{Z}_1$, $\mathbf{X}$, $\mathbf{Y}$, $\mathbf{Z}$ will be such that no part of them is periodic in any of the periods of the orbits of the charges. Hence, in calculating $\mathbf{X}$, $\mathbf{Y}$, $\mathbf{Z}$ from the known expressions for the electric forces acting on a single charge, only those terms which are not periodic in the periods of the orbits of the charges need be taken account of, and the conditions of permanence of the group can be obtained from the relations which have to be satisfied in order that the periodic terms should vanish.

Taking first the case of moving charges describing orbits of small linear dimensions about fixed points, the equations (20) become

$$\begin{aligned} \mathbf{X}_1 + \mathbf{X} &= 0, \\ \mathbf{Y}_1 + \mathbf{Y} &= 0, \\ \mathbf{Z}_1 + \mathbf{Z} &= 0, \end{aligned} \right\} \quad \dots \dots \dots \dots \dots (21).$$

Assuming that radiation is wholly an electrical effect, the components X_1, Y_1, Z_1 of the non-electrical forces acting on a single charge will contain terms which are periodic in time and have periods depending on the periods of the orbits of the charges, and therefore, in effecting the summations $\Sigma e X'$, $\Sigma e Y'$, $\Sigma e Z'$ for the group, the periodic terms must disappear.

Now
$$X' = X + \dot{\eta}\gamma - \dot{\zeta}\beta,$$
$$Y' = Y + \dot{\zeta}\alpha - \dot{\xi}\gamma,$$
$$Z' = Z + \dot{\xi}\beta - \dot{\eta}\alpha,$$

where $\dot{\xi}, \dot{\eta}, \dot{\zeta}$ are the velocities of the charge, and $X, Y, Z, \alpha, \beta, \gamma$ are the components of the electric and magnetic forces at the point occupied by the charge, therefore the periodic terms which occur arise from terms of the type ξ, ξ^2, $\eta\zeta$, ..., $\dot{\xi}$, $\dot{\xi}^2$, $\eta\dot{\zeta}$, ..., and the conditions that the periodic terms should disappear from $\Sigma eX'$, $\Sigma eY'$, $\Sigma eZ'$ are such that the non-periodic terms which involve the dimensions of the orbits also disappear. Hence, if P, Q, R denote the components of the force of electrical origin which one permanent group exerts on another, and

$$P = P_1 + P_2, \quad Q = Q_1 + Q_2, \quad R = R_1 + R_2,$$

where P_2, Q_2, R_2 are the parts of P, Q, R which involve the dimensions of the orbits, the forces ΣP_2, ΣQ_2, ΣR_2 which all the other groups exert on any one group vanish, and the equations (21) become

$$\left.\begin{aligned}
\mathbf{X}_1 + \Sigma P_1 &= 0, \\
\mathbf{Y}_1 + \Sigma Q_1 &= 0, \\
\mathbf{Z}_1 + \Sigma R_1 &= 0,
\end{aligned}\right\} \quad \ldots\ldots\ldots\ldots\ldots\ldots (22).$$

The forces P_1, Q_1, R_1, being those which are independent of the dimensions of the orbits, are given by

$$\left.\begin{aligned}
P_1 &= -\Sigma e V^2 \frac{\partial}{\partial x} \Sigma \frac{e'}{r}, \\
Q_1 &= -\Sigma e V^2 \frac{\partial}{\partial y} \Sigma \frac{e'}{r}, \\
R_1 &= -\Sigma e V^2 \frac{\partial}{\partial z} \Sigma \frac{e'}{r},
\end{aligned}\right\} \quad \ldots\ldots\ldots\ldots\ldots (23),$$

in which the first Σ refers to the group which is being considered, the second Σ to any other group, and r denotes the distance AB, where A is one of the fixed points about which a charge belonging to the first group is describing its orbit and B is such a point belonging to the second group. When there is no free electricity, the condition $\Sigma e = 0$ will be satisfied for each group, and the forces P_1, Q_1, R_1 are of the order $V^2 e^2 d^2 / r^4$, where r is the distance between two groups and d is a length depending on the distances between the orbits of the same group. When there is free electricity the forces on a group for which Σe is not zero are those of electrostatic theory. In every case the forces P_2, Q_2, R_2, are of the order $V^2 e^2 l^2 d^2 / \lambda^2 r^4$, where l is a length depending on the dimensions of an orbit and λ is the wave length in the aether corresponding to a period in the orbit; thus the forces P_2, Q_2, R_2 are small compared with the forces P_1, Q_1, R_1*.

It appears from the preceding that, when the groups and the distances between the orbits in the groups are arranged so that the forces P_1, Q_1, R_1 form with the non-electrical forces an equilibrating system, the orbital motions will arrange themselves so that the forces P_2, Q_2, R_2 form an equilibrating system, as ΣP_2, ΣQ_2, ΣR_2 vanish when the conditions of permanence of the groups are satisfied. Hence, in investigating the law of distribution of the groups, it is sufficient to consider the forces P_1, Q_1, R_1 and the non-electrical forces. If now it be assumed that the non-electrical forces† are such that they are not sensible at insensible distances, which is equivalent to assuming that the inter-atomic forces are wholly electrical, the forces P_1, Q_1, R_1 will form an equilibrating system when there is no free electricity, and if, when there is free electricity, P_1', Q_1', R_1', denote the parts of P_1, Q_1, R_1 which remain after the removal of the forces belonging to the free electricity, the forces P_1', Q_1', R_1' will form an equilibrating system. Therefore, when this assumption is made, a knowledge of the forces P_1, Q_1, R_1 will suffice for the determination of the law of distribution of the groups.

Considering now the case where the points, about which the charges are describing their orbits, are all moving with a uniform velocity v in the direction of the axis of x, the equations (20) become, since v is constant,

$$\mathbf{X}_1 + \mathbf{X} = 0,$$
$$\mathbf{Y}_1 + \mathbf{Y} = 0,$$
$$\mathbf{Z}_1 + \mathbf{Z} = 0,$$

the same equations as in the previous case. The same arguments will apply as in that case, and the law of distribution

---

* Cf. §§ 53, 54.

† The force of gravitation satisfies the condition.

of the groups can be determined from a knowledge of the forces P_1, Q_1, R_1. By a preceding investigation* the values of P_1, Q_1, R_1 for this case are given by

$$P_1 = -\Sigma e V^2 (1 - \kappa^2) \frac{\partial}{\partial x} \Sigma \frac{e'}{R \sqrt{(1 - \kappa^2)}},$$

$$Q_1 = -\Sigma e V^2 \frac{\partial}{\partial y} \Sigma \frac{e'}{R \sqrt{(1 - \kappa^2)}}, \quad \Bigg\} \quad \ldots\ldots(24),$$

$$R_1 = -\Sigma e V^2 \frac{\partial}{\partial z} \Sigma \frac{e'}{R \sqrt{(1 - \kappa^2)}},$$

where $\kappa = \mathrm{v}/V$, and

$$R^2 = (x - x_1)^2/(1 - \kappa^2) + (y - y_1)^2 + (z - z_1)^2,$$

x, y, z being the coordinates of the point about which a charge in the first group is describing its orbit and x_1, y_1, z_1 a similar point for the second group. Writing

$$x = x' \sqrt{(1 - \kappa^2)},$$

the equations (24) become

$$P_1 = -\sqrt{(1 - \kappa^2)} \Sigma e V^2 \frac{\partial}{\partial x'} \Sigma \frac{e'}{R' \sqrt{(1 - \kappa^2)}},$$

$$Q_1 = -\Sigma e V^2 \frac{\partial}{\partial y} \Sigma \frac{e'}{R' \sqrt{(1 - \kappa^2)}},$$

$$R_1 = -\Sigma e V^2 \frac{\partial}{\partial z} \Sigma \frac{e'}{R' \sqrt{(1 - \kappa^2)}},$$

where

$$R'^2 = (x' - x_1')^2 + (y - y_1)^2 + (z - z_1)^2,$$

and putting $P_1 = P_1' \sqrt{1 - \kappa^2}$, it being remembered that $\ddot{x} = \ddot{x}' \sqrt{(1 - \kappa^2)}$, these equations become

$$P_1' = -\Sigma e V^2 \frac{\partial}{\partial x'} \Sigma \frac{e'}{R' \sqrt{(1 - \kappa^2)}},$$

$$Q_1 = -\Sigma e V^2 \frac{\partial}{\partial y} \Sigma \frac{e'}{R' \sqrt{(1 - \kappa^2)}}, \quad \Bigg\} \quad \ldots\ldots\ldots(25).$$

$$R_1 = -\Sigma e V^2 \frac{\partial}{\partial z} \Sigma \frac{e'}{R' \sqrt{(1 - \kappa^2)}},$$

* p. 177.

Now the equations (25) give the values of P_1', Q_1, R_1 which belong to a distribution, in which the points about which the orbits are being described are at rest; therefore, if this system is in equilibrium, the groups which form the system to which the equations (24) belong will also be in a state of relative equilibrium. Hence the relation between the configuration of the groups, when moving with a uniform velocity v, and, when they are at rest is given by the equation $x = x' \sqrt{(1 - \kappa^2)}$, where $\kappa = v/V$; that is, the material medium, which consists of these groups, is contracted in the direction of its motion in the ratio $\sqrt{(1 - v^2/V^2)} : 1$*.

In the investigation it has been assumed that there is no electric charge which is migrating freely among the groups, so that the result will not necessarily apply to the case of a material medium in which there are electric currents. It may be shewn by similar reasoning that the result holds for a material medium in which there are no electric currents, when waves due to external causes are being propagated through it, provided none of the periods of these waves are the same as those belonging to the orbits of the charges which form the groups. The medium would then, if electrically isotropic when at rest, not be so when in motion, the specific inductive capacity of the medium in the direction of the motion being altered by an amount depending on κ^2 and higher powers of κ; and to obtain the effect of the motion of a material medium on the velocity of propagation of waves through it, when κ^2 is not negligible, this alteration would have to be taken into account.

---

* This result has been obtained for the case of charges moving with a uniform velocity by Lorentz and Larmor.

APPENDIX D.

DIFFRACTION.

WHEN electric waves are incident on any body, other waves due to the presence of the body are set up, and to determine completely the superposed system of waves it would be necessary to know the distribution of electric charges which constitute the body; the motion would then be expressed in terms of coordinates specifying the degrees of freedom of these electric charges and the electrical degrees of freedom of the aether. In the treatment of any class of cases it is convenient to suppose that the effects of the motions of the electric charges can be expressed in terms of a smaller number of coordinates, or be represented by kinematical conditions. In the case where the body is transparent these effects are suitably represented by two coordinates for each point of the body, these coordinates being specified by a vector f_1, g_1, h_1, which satisfies the relation

$$\frac{\partial f_1}{\partial x} + \frac{\partial g_1}{\partial y} + \frac{\partial h_1}{\partial z} = 0$$

at every point of the body; this vector is termed the material electric displacement. In the case of bodies which are not transparent this representation is not suitable, but two limiting cases can be recognized, which will furnish a clue to the effect on the waves of any such body. The first extreme supposition is that the electric charges at the surface of the body move in such a way that waves of the same kind and of the same amount are radiated out from them as are incident on the surface of the body, so that no part of the energy of the waves in the medium is absorbed by the body ; this is the case in which the body is a perfect conductor. The other extreme supposition is that all the energy of the waves incident on the body is absorbed by it.

The problem of the diffraction of waves by a transparent body has been solved for the case of a circular cylinder* and for that of a sphere†, the velocity of radiation in the body differing by a finite amount from that in the surrounding medium. The problem has been solved for the general case, when the difference between the velocities of radiation is very small‡. The problem of the diffraction of waves by a perfectly conducting body has been solved for the case of a circular cylinder§, a sphere‖, and an indefinitely thin wedge in the form of a semi-infinite plane¶. When the body is perfectly absorbing the problem can always be solved. One of the most important cases is that in which the diffracting body consists of a plane screen with an aperture in it. This case has been treated in detail by Stokes** and by Lorenz ††, who have assumed that the secondary or diffracted waves depend only on the disturbance over the aperture ; this assumption is equivalent to assuming that the screen is a perfect absorber. If the problem could be solved for a perfectly conducting screen, the comparison of the two results would shew in what respect either representation is ineffective and possibly give information as to where improvement might be made. The problem has so far only been solved for the case of the semi-infinite plane, and its importance makes it desirable to have a solution as direct and simple as possible. The application of the methods used in the ninth chapter of this essay leads directly and easily to

* Lord Rayleigh, *Phil. Mag.* xii. 1881.

† Lorenz, *Vidensk. Selsk. Skr.* Copenhagen, 1890.

‡ Lord Rayleigh, *l.c.*

§ J. J. Thomson, *Recent Researches in Electricity and Magnetism*, p. 428. 1893.

‖ J. J. Thomson, *l.c.* p. 437.

¶ Poincaré, *Acta Mathematica*, Vol. xvi. 1892–3, Vol. xx. 1897. Sommerfeld, *Math. Ann.* Vol. xlvii. 1896.

** *Camb. Phil. Trans.* Vol. ix. 1849.

†† *Pogg. Ann.* cxi. 1860 ; *Crelle,* lviii. 1861.

the solution of a more general case, viz. that in which the diffracting body is a wedge of any angle.

Let the edge of the wedge be chosen as the axis of z, and let r, z, θ be cylindrical coordinates of a point so that the faces of the wedge are given by $\theta = 0$, $\theta = \alpha$, and the space occupied by it is that between $\theta = \alpha$ and $\theta = 2\pi$. Taking first the case in which the electric force is parallel to the edge of the wedge, the differential equation satisfied by Z at all points at which there are no sources is

$$\frac{\partial^2 Z}{\partial r^2} + \frac{1}{r}\frac{\partial Z}{\partial r} + \frac{1}{r^2}\frac{\partial^2 Z}{\partial \theta^2} = \frac{1}{V^2}\frac{\partial^2 Z}{\partial t^2},$$

that is, for waves of wave length $2\pi/\kappa$,

$$\frac{\partial^2 Z}{\partial r^2} + \frac{1}{r}\frac{\partial Z}{\partial r} + \frac{1}{r^2}\frac{\partial^2 Z}{\partial \theta^2} + \kappa^2 Z = 0 \quad\ldots\ldots\ldots\ldots(1).$$

For waves of the kind considered the sources can be represented by lines of discontinuity of electric force parallel to the axis of z and at points on such a line Z will satisfy the differential equation

$$\frac{\partial^2 Z}{\partial r^2} + \frac{1}{r}\frac{\partial Z}{\partial r} + \frac{1}{r^2}\frac{\partial^2 Z}{\partial \theta^2} + \kappa^2 Z + 2\pi\rho = 0 \quad\ldots\ldots\ldots(2).$$

The condition to be satisfied at the surface of the wedge is that Z vanishes when $\theta = 0$ and when $\theta = \alpha$. The solution of (2) is therefore

$$Z = \sum_1^\infty R_n \sin \frac{n\pi\theta}{\alpha},$$

where n has all positive integral values and R_n is a function of r to be determined from the relation

$$\sum_1^\infty \left\{ \frac{\partial^2 R_n}{\partial r^2} + \frac{1}{r}\frac{\partial R_n}{\partial r} + \left(\kappa^2 - \frac{n^2\pi^2}{\alpha^2 r^2}\right) R_n \right\} \sin \frac{n\pi\theta}{\alpha} + 2\pi\rho = 0 \,;$$

whence R_n satisfies the equation

$$\frac{\partial^2 R_n}{\partial r^2} + \frac{1}{r}\frac{\partial R_n}{\partial r} + \left(\kappa^2 - \frac{n^2\pi^2}{\alpha^2 r^2}\right) R_n + \frac{4\pi}{\alpha}\int_0^\alpha \rho \sin\frac{n\pi\theta'}{\alpha} d\theta' = 0 \ldots(3).$$

It is sufficient to consider the case in which there is a single line of sources determined by $r = r_1$, $\theta = \theta_1$; the solution of (3) is given by

$$R_n = A_n J_{\frac{n\pi}{a}}(\kappa r),$$

when $r < r_1$, for R_n must be finite when $r = 0$, and by

$$R_n = B_n \{ J_{-\frac{n\pi}{a}}(\kappa r) - e^{\frac{n\pi^2 \iota}{a}} J_{\frac{n\pi}{a}}(\kappa r) \},$$

when $r > r_1$, for R_n cannot involve $e^{\iota \kappa r}$, there being no reflexion at the infinitely distant boundary. Hence

$$Z = \sum_{1}^{\infty} A_n J_{\frac{n\pi}{a}}(\kappa r) \sin \frac{n\pi\theta}{\alpha},$$

when $r < r_1$, and

$$Z = \sum_{1}^{\infty} B_n \{ J_{-\frac{n\pi}{a}}(\kappa r) - e^{\frac{n\pi^2 \iota}{a}} J_{\frac{n\pi}{a}}(\kappa r) \} \sin \frac{n\pi\theta}{\alpha},$$

when $r > r_1$; now both these series will converge and be identical, when $r = r_1$, except when $\theta = \theta_1$, in which case they will diverge; therefore

$$A_n J_{\frac{n\pi}{a}}(\kappa r_1) = B_n \{ J_{-\frac{n\pi}{a}}(\kappa r_1) - e^{\frac{n\pi^2 \iota}{a}} J_{\frac{n\pi}{a}}(\kappa r_1) \},$$

and the solution can be written

$$Z = \sum_{1}^{\infty} C_n J_{\frac{n\pi}{a}}(\kappa r) \{ J_{-\frac{n\pi}{a}}(\kappa r_1) - e^{\frac{n\pi^2 \iota}{a}} J_{\frac{n\pi}{a}}(\kappa r_1) \} \sin \frac{n\pi\theta}{\alpha},$$

when $r < r_1$, and

$$Z = \sum_{1}^{\infty} C_n J_{\frac{n\pi}{a}}(\kappa r_1) \{ J_{-\frac{n\pi}{a}}(\kappa r) - e^{\frac{n\pi^2 \iota}{a}} J_{\frac{n\pi}{a}}(\kappa r) \} \sin \frac{n\pi\theta}{\alpha},$$

when $r > r_1$. It appears at once, as in § 60, by considering the case where $\alpha = \pi$ which corresponds to reflexion at a plane surface, that C_n is independent of κ, and it is therefore sufficient to determine C_n for the case in which $\kappa = 0$. In this case the problem to be solved is the electrostatic one of a line charge influencing a conducting wedge, and the potential due to a line charge of strength m is known to be

$$\frac{m}{2} \log \frac{r^{\frac{2\pi}{a}} + r_1^{\frac{2\pi}{a}} - 2 (rr_1)^{\frac{\pi}{a}} \cos \frac{\pi}{\alpha}(\theta + \theta_1)}{r^{\frac{2\pi}{a}} + r_1^{\frac{2\pi}{a}} - 2 (rr_1)^{\frac{\pi}{a}} \cos \frac{\pi}{\alpha}(\theta - \theta_1)}.$$

Therefore, when $\kappa = 0$,

$$Z = 2m \sum_1^\infty \frac{1}{n} \left(\frac{r}{r_1}\right)^{\frac{n\pi}{a}} \sin\frac{n\pi\theta}{\alpha} \sin\frac{n\pi\theta_1}{\alpha},$$

when $r < r_1$, and

$$Z = 2m \sum_1^\infty \frac{1}{n} \left(\frac{r_1}{r}\right)^{\frac{n\pi}{a}} \sin\frac{n\pi\theta}{\alpha} \sin\frac{n\pi\theta_1}{\alpha},$$

when $r > r_1$, whence

$$C_n \operatorname{Lt}_{\kappa=0} J^{n\pi}_{\frac{}{a}}(\kappa r) \left\{ J_{-\frac{n\pi}{a}}(\kappa r_1) - e^{\frac{n\pi^2 \iota}{a}} J_{\frac{n\pi}{a}}(\kappa r_1) \right\} = \frac{2m}{n}\left(\frac{r}{r_1}\right)^{\frac{n\pi}{a}} \sin\frac{n\pi\theta_1}{\alpha},$$

that is

$$C_n \frac{r^{\frac{n\pi}{a}}}{2^{\frac{n\pi}{a}} \, \Pi\left(\frac{n\pi}{\alpha}\right)} \cdot \frac{r_1^{-\frac{n\pi}{a}}}{2^{-\frac{n\pi}{a}} \, \Pi\left(-\frac{n\pi}{\alpha}\right)} = \frac{2m}{n}\left(\frac{r}{r_1}\right)^{\frac{n\pi}{a}} \sin\frac{n\pi\theta_1}{\alpha},$$

and therefore

$$C_n = \frac{2\pi m}{\alpha} \cdot \frac{\pi}{\sin\frac{n\pi^2}{\alpha}} \sin\frac{n\pi\theta_1}{\alpha}.$$

Hence, when there is a line of electric discontinuity of strength m at $r = r_1$, $\theta = \theta_1$, the electric force is given by

$$Z = \frac{2\pi m}{\alpha} \sum_1^\infty J_{\frac{n\pi}{a}}(\kappa r)\left\{ J_{-\frac{n\pi}{a}}(\kappa r_1) - e^{\frac{n\pi^2 \iota}{a}} J_{\frac{n\pi}{a}}(\kappa r_1) \right\} \frac{\pi}{\sin\frac{n\pi^2}{\alpha}}$$
$$. \sin\frac{n\pi\theta}{\alpha} \sin\frac{n\pi\theta_1}{\alpha},$$

when $r < r_1$, and by

$$Z = \frac{2\pi m}{\alpha} \sum_1^\infty J_{\frac{n\pi}{a}}(\kappa r_1)\left\{ J_{-\frac{n\pi}{a}}(\kappa r) - e^{\frac{n\pi^2 \iota}{a}} J_{\frac{n\pi}{a}}(\kappa r) \right\} \frac{\pi}{\sin\frac{n\pi^2}{\alpha}}$$
$$. \sin\frac{n\pi\theta}{\alpha} \sin\frac{n\pi\theta_1}{\alpha},$$

when $r > r_1$. Now

$$J_{\frac{n\pi}{a}}(\kappa r)\left\{ J_{-\frac{n\pi}{a}}(\kappa r_1) - e^{\frac{n\pi^2 \iota}{a}} J_{\frac{n\pi}{a}}(\kappa r_1) \right\} \frac{\pi}{\sin\frac{n\pi^2}{\alpha}}$$
$$= -\int_{c-\infty\iota}^0 e^{\frac{t}{2} - \frac{\kappa^2(r^2 + r_1^2)}{2t}} I_{\frac{n\pi}{a}}\left(\frac{\kappa^2 r r_1}{t}\right) \frac{dt}{t},*$$

* *Proc. Lond. Math. Soc.* Vol. XXXII. p. 155, 1900.

where c is a real positive quantity, therefore

$$Z = -\frac{2\pi m}{\alpha} \sum_1^\infty \int_{c-\infty\iota}^0 e^{\frac{t}{2}-\frac{\kappa^2(r^2+r_1^2)}{2t}} I_{\frac{n\pi}{a}}\left(\frac{\kappa^2 r r_1}{t}\right) \frac{dt}{t} \sin\frac{n\pi\theta}{\alpha} \sin\frac{n\pi\theta_1}{\alpha}$$

for all values of r. Again

$$I_{\frac{n\pi}{a}}\left(\frac{\kappa^2 r r_1}{t}\right) = \frac{1}{2\pi\iota} \int_{\infty e^{\gamma\iota}}^{\infty e^{\gamma'\iota}} e^{\frac{\kappa^2 r r_1}{2t}\left(s+\frac{1}{s}\right)} \frac{ds}{s^{\frac{n\pi}{a}+1}}, *$$

where the real parts of $\infty\, e^{\gamma\iota}/t$, $\infty\, e^{\gamma'\iota}/t$ are negative, whence, writing $s = e^{-\iota\varsigma}$, where $\varsigma = \xi + \iota\eta$,

$$I_{\frac{n\pi}{a}}\left(\frac{\kappa^2 r r_1}{t}\right) = -\frac{1}{2\pi} \int_{\infty\iota+\gamma}^{\infty\iota+\gamma'} e^{\frac{\kappa^2 r r_1}{t}\cos\varsigma + \frac{n\pi\varsigma\iota}{a}}\, d\varsigma,$$

where $2\pi > \gamma > \pi$, $0 > \gamma' > -\pi$, and therefore

$$Z = \frac{m}{\alpha} \sum_1^\infty \int_{c-\infty\iota}^0 \int_{\infty\iota+\gamma}^{\infty\iota+\gamma'} e^{\frac{t}{2}-\frac{\kappa^2}{2t}(r^2+r_1^2-2rr_1\cos\varsigma)+\frac{n\pi\varsigma\iota}{a}} \frac{dt}{t}\, d\varsigma$$

$$. \sin\frac{n\pi\theta}{\alpha} \sin\frac{n\pi\theta_1}{\alpha},$$

that is

$$Z = -\frac{m\iota}{4\alpha} \int_{c-\infty\iota}^0 \int_{\infty\iota+\gamma}^{\infty\iota+\gamma'} e^{\frac{t}{2}-\frac{\kappa^2}{2t}(r^2+r_1^2-2rr_1\cos\varsigma)} \frac{dt}{t} \left\{ \frac{\sin\frac{\pi\varsigma}{\alpha}}{\cos\frac{\pi\varsigma}{\alpha}-\cos\frac{\pi}{\alpha}(\theta-\theta_1)} \right.$$

$$\left. - \frac{\sin\frac{\pi\varsigma}{\alpha}}{\cos\frac{\pi\varsigma}{\alpha}-\cos\frac{\pi}{\alpha}(\theta+\theta_1)} \right\} d\varsigma,$$

where the path of integration lies wholly in the upper half of the ς plane. Changing the order of integration and using the known result

$$K_0(\iota z) = -\tfrac{1}{2} \int_{c-\infty\iota}^0 e^{\frac{t}{2}-\frac{z^2}{2t}} \frac{dt}{t},$$

* *Proc. Lond. Math. Soc.* Vol. xxx. p. 167, 1899.

the above becomes

$$Z = \frac{m\iota}{2a} \int_{\infty\iota+\gamma}^{\infty\iota+\gamma'} K_0 \left\{ \iota\kappa \sqrt{(r^2 + r_1^2 - 2rr_1 \cos \zeta)} \right\} \left\{ \frac{\sin \dfrac{\pi\zeta}{\alpha}}{\cos \dfrac{\pi\zeta}{\alpha} - \cos \dfrac{\pi}{\alpha}(\theta - \theta_1)} - \frac{\sin \dfrac{\pi\zeta}{\alpha}}{\cos \dfrac{\pi\zeta}{\alpha} - \cos \dfrac{\pi}{\alpha}(\theta + \theta_1)} \right\} d\zeta.$$

When α is an integral submultiple of π, $\alpha = \pi/n$, this is equivalent to

$$Z = m \sum_{s=0}^{n-1} K_0 \left[\iota\kappa \sqrt{\left\{ r^2 + r_1^2 - 2rr_1 \cos\left(\theta - \theta_1 - \frac{2s\pi}{n}\right) \right\}} \right]$$
$$- m \sum_{s=0}^{n-1} K_0 \left[\iota\kappa \sqrt{\left\{ r^2 + r_1^2 - 2rr_1 \cos\left(\theta + \theta_1 + \frac{2s\pi}{n}\right) \right\}} \right],$$

which, remembering that the effect of a line of electric discontinuity of strength m in an unbounded space is $mK_0(\iota\kappa R)$, where R is the distance of any point from this line, is the result which would be arrived at by the method of images. When α is an integral multiple of π, the above expression becomes identical with that given by Sommerfeld* for this case.

The solution in the case of plane waves can be obtained from this by making r_1 indefinitely great. The value to which $K_0 \{ \iota\kappa \sqrt{(r^2 + r_1^2 - 2rr_1 \cos \zeta)} \}$ tends is

$$- \frac{\pi\iota}{\sqrt{(2\pi\kappa r_1)}} e^{\frac{\pi\iota}{4} - \iota\kappa r_1 + \iota\kappa r \cos \zeta};$$

hence, corresponding to plane waves incident in the direction θ_1 and in which the electric force is given by $e^{\iota\kappa r \cos(\theta - \theta_1)}$, the solution for the space bounded by the planes $\theta = 0$, $\theta = \alpha$ is

$$Z = \frac{\iota}{2a} \int_{\infty\iota+\gamma}^{\infty\iota+\gamma'} e^{\iota\kappa r \cos \zeta} \left\{ \frac{\sin \dfrac{\pi\zeta}{\alpha}}{\cos \dfrac{\pi\zeta}{\alpha} - \cos \dfrac{\pi}{\alpha}(\theta - \theta_1)} - \frac{\sin \dfrac{\pi\zeta}{\alpha}}{\cos \dfrac{\pi\zeta}{\alpha} - \cos \dfrac{\pi}{\alpha}(\theta + \theta_1)} \right\} d\zeta,$$

where $2\pi > \gamma > \pi$, $0 > \gamma' > -\pi$.

* l.c. p. 187.

When $\alpha = 2\pi$, the wedge becomes a semi-infinite plane and the above expression for Z becomes

$$Z = \frac{\iota}{4\pi} \int_{\infty\iota+\gamma}^{\infty\iota+\gamma'} e^{\iota\kappa r \cos\zeta}$$

$$\cdots \left\{ \frac{\sin\dfrac{\zeta}{2}}{\cos\dfrac{\zeta}{2} - \cos\tfrac{1}{2}(\theta - \theta_1)} - \frac{\sin\dfrac{\zeta}{2}}{\cos\dfrac{\zeta}{2} - \cos\tfrac{1}{2}(\theta + \theta_1)} \right\} d\zeta.$$

The integrals can now be transformed as follows: writing

$$I = \frac{\iota}{4\pi} \int_{\infty\iota+\gamma}^{\infty\iota+\gamma'} e^{\iota\kappa r \cos\zeta} \frac{\sin\dfrac{\zeta}{2}}{\cos\dfrac{\zeta}{2} - \cos\tfrac{1}{2}(\theta - \theta_1)} d\zeta,$$

and putting $\cos\dfrac{\zeta}{2} = t \cos\tfrac{1}{2}(\theta - \theta_1)$,

$$I = -\frac{\iota}{2\pi} e^{\iota\kappa r \cos(\theta - \theta_1)} \int_{t_0}^{t_1} e^{2\iota\kappa r \cos^2\frac{1}{2}(\theta - \theta_1)(t^2-1)} \frac{dt}{t-1};$$

where, if $\cos\tfrac{1}{2}(\theta - \theta_1)$ is positive,

$$t_0 = -c - \infty\iota, \quad t_1 = c + \infty\iota,$$

the path of integration cutting the real axis in the t plane to the right of the point $t = 1$, and, if $\cos\tfrac{1}{2}(\theta - \theta_1)$ is negative,

$$t_0 = c + \infty\iota, \quad t_1 = -c - \infty\iota,$$

the path of integration cutting the real axis in the t plane to the left of the origin. Hence, deforming the path so that it becomes the imaginary axis in the t plane,

$$I = e^{\iota\kappa r \cos(\theta - \theta_1)} - \frac{\iota}{2\pi} e^{\iota\kappa r \cos(\theta - \theta_1)} \int_{-\infty\iota}^{\infty\iota} e^{2\iota\kappa r \cos^2\frac{1}{2}(\theta - \theta_1)(t^2-1)} \frac{dt}{t-1},$$

when $\cos\tfrac{1}{2}(\theta - \theta_1)$ is positive, and

$$I = \frac{\iota}{2\pi} e^{\iota\kappa r \cos(\theta - \theta_1)} \int_{-\infty\iota}^{\infty\iota} e^{2\iota\kappa r \cos^2\frac{1}{2}(\theta - \theta_1)(t^2-1)} \frac{dt}{t-1},$$

when $\cos\tfrac{1}{2}(\theta - \theta_1)$ is negative. Denoting the integral in the above relations by I_1, and writing $\mu_0 = \sqrt{(2\kappa r)}|\cos\tfrac{1}{2}(\theta - \theta_1)|$, then

$$I_1 = \frac{\iota}{2\pi} e^{\iota\kappa r \cos(\theta - \theta_1)} \int_{-\infty\iota}^{\infty\iota} e^{\iota\mu_0^2(t^2-1)} \frac{dt}{t-1},$$

that is,

$$I_1 = \frac{\iota}{2\pi} e^{\iota \kappa r \cos(\theta - \theta_1)} \int_0^{\infty \iota} e^{\iota \mu_0^2 (t^2 - 1)} \left\{ \frac{1}{t-1} - \frac{1}{t+1} \right\} dt,$$

which becomes

$$I_1 = \frac{\iota}{\pi} e^{\iota \kappa r \cos(\theta - \theta_1)} \int_0^{\infty \iota} e^{\iota \mu_0^2 (t^2 - 1)} \frac{dt}{t^2 - 1},$$

or

$$I_1 = -\frac{1}{\pi} e^{\iota \kappa r \cos(\theta - \theta_1)} \int_0^{\infty} e^{-\iota \mu_0^2 (t^2 + 1)} \frac{dt}{t^2 + 1}.$$

Now

$$\frac{1}{1 + t^2} = \int_0^{\infty} e^{-u(1 + t^2)} du,$$

therefore

$$I_1 = -\frac{1}{\pi} e^{\iota \kappa r \cos(\theta - \theta_1)} \int_0^{\infty} \int_0^{\infty} e^{-(u + \iota \mu_0^2)(1 + t^2)} dt,$$

which, changing the order of integration and integrating with respect to t, is equivalent to

$$I_1 = -\frac{1}{2\sqrt{\pi}} e^{\iota \kappa r \cos(\theta - \theta_1)} \int_0^{\infty} e^{-(u + \iota \mu_0^2)} \frac{du}{\sqrt{(u + \iota \mu_0^2)}},$$

and, putting $u + \iota \mu_0^2 = \iota v^2$, this becomes

$$I_1 = -\frac{1}{2\sqrt{\pi}} e^{\iota \kappa r \cos(\theta - \theta_1) + \frac{\pi \iota}{4}} \int_{\mu_0 e^{\frac{\pi \iota}{4}}}^{\infty e^{\frac{3\pi \iota}{4}}} e^{-\iota v^2} dv,$$

that is

$$I_1 = \frac{1}{\sqrt{\pi}} e^{\iota \kappa r \cos(\theta - \theta_1) + \frac{\pi \iota}{4}} \int_{\mu_0}^{\infty} e^{-\iota v^2} dv.$$

Hence

$$I = e^{\iota \kappa r \cos(\theta - \theta_1)} \left\{ 1 - \frac{e^{\frac{\pi \iota}{4}}}{\sqrt{\pi}} \int_{\mu_0}^{\infty} e^{-\iota v^2} dv \right\},$$

when $\cos \frac{1}{2}(\theta - \theta_1)$ is positive, that is

$$I = \frac{1}{\sqrt{\pi}} e^{\iota \kappa r \cos(\theta - \theta_1) + \frac{\pi \iota}{4}} \left\{ \int_{-\infty}^{\infty} e^{-\iota v^2} dv - \int_{\mu_0}^{\infty} e^{-\iota v^2} dv \right\},$$

or

$$I = \frac{1}{\sqrt{\pi}} e^{\iota \kappa r \cos(\theta - \theta_1) + \frac{\pi \iota}{4}} \int_{-\infty}^{\mu_0} e^{-\iota v^2} dv.$$

Also

$$I = \frac{1}{\sqrt{\pi}} e^{\iota \kappa r \cos(\theta - \theta_1) + \frac{\pi \iota}{4}} \int_{\mu_0}^{\infty} e^{-\iota v^2} dv,$$

when $\cos \frac{1}{2}(\theta - \theta_1)$ is negative, that is, in this case,

$$I = \frac{1}{\sqrt{\pi}} e^{\iota \kappa r \cos(\theta - \theta_1) + \frac{\pi \iota}{4}} \int_{-\infty}^{-\mu_0} e^{-\iota v^2} dv.$$

Therefore

$$I = \frac{1}{\sqrt{\pi}} e^{\iota \kappa r \cos(\theta - \theta_1) + \frac{\pi \iota}{4}} \int_{-\infty}^{\mu} e^{-\iota v^2} dv,$$

where $\mu = \sqrt{(2\kappa r)} \cos \frac{1}{2}(\theta - \theta_1)$, and the electric force Z is given by

$$Z = \frac{1}{\sqrt{\pi}} e^{\iota \kappa r \cos(\theta - \theta_1) + \frac{\pi \iota}{4}} \int_{-\infty}^{\mu} e^{-\iota v^2} dv - \frac{1}{\sqrt{\pi}} e^{\iota \kappa r \cos(\theta + \theta_1) + \frac{\pi \iota}{4}} \int_{-\infty}^{\mu'} e^{-\iota v^2} dv,$$

where $\mu' = \sqrt{(2\kappa r)} \cos \frac{1}{2}(\theta + \theta_1)$.

When the electric force is in the plane of incidence of the waves, the magnetic induction is parallel to the edge of the wedge, and denoting the magnetic induction by c, c will satisfy the same equation as Z above, the boundary conditions now being $\frac{\partial c}{\partial \theta} = 0$, when $\theta = 0$ and when $\theta = \alpha$. Thus

$$c = \sum_0^\infty R_n \cos \frac{n\pi\theta}{\alpha},$$

which, as before, becomes

$$c = \sum_0^\infty C_n J_{\frac{n\pi}{\alpha}}(\kappa r) \{ J_{-\frac{n\pi}{\alpha}}(\kappa r_1) - e^{-\frac{n\pi^2 \iota}{\alpha}} J_{\frac{n\pi}{\alpha}}(\kappa r_1)\} \cos \frac{n\pi\theta}{\alpha},$$

when $r < r_1$, and

$$c = \sum_0^\infty C_n J_{\frac{n\pi}{\alpha}}(\kappa r_1) \{ J_{-\frac{n\pi}{\alpha}}(\kappa r) - e^{-\frac{n\pi^2 \iota}{\alpha}} J_{\frac{n\pi}{\alpha}}(\kappa r)\} \cos \frac{n\pi\theta}{\alpha},$$

when $r > r_1$, and again C_n is independent of κ. When $\kappa = 0$, the problem is the hydrodynamical one of a line source in the space bounded by $\theta = 0$, and $\theta = \alpha$. The solution is then the same as in the previous case, with the exception that the part involving $\theta + \theta_1$ is affected with the positive instead of the negative sign, and therefore

$$c = \frac{m\iota}{2\alpha} \int_{\infty \iota + \gamma}^{\infty \iota + \gamma} K_0 \{ \iota \kappa \sqrt{(r^2 + r_1^2 - 2rr_1 \cos \zeta)} \}$$

$$\cdot \left\{ \frac{\sin \frac{\pi \zeta}{\alpha}}{\cos \frac{\pi \zeta}{\alpha} - \cos \frac{\pi}{\alpha}(\theta - \theta_1)} + \frac{\sin \frac{\pi \zeta}{\alpha}}{\cos \frac{\pi \zeta}{\alpha} - \cos \frac{\pi}{\alpha}(\theta + \theta_1)} \right\} d\zeta.$$

For plane waves incident on a semi-infinite plane this becomes

$$c = \frac{1}{\sqrt{\pi}} e^{\iota \kappa r \cos(\theta - \theta_1) + \frac{\pi \iota}{4}} \int_{-\infty}^{\mu} e^{-\iota v^2} dv + \frac{1}{\sqrt{\pi}} e^{\iota \kappa r \cos(\theta + \theta_1) + \frac{\pi \iota}{4}} \int_{-\infty}^{\mu'} e^{-\iota v^2} dv.$$

The problem solved by Poincaré in the first instance was that of a wedge-shaped beam converging on the edge of the semi-infinite plane; this can be obtained from the previous general expression. The comparison* of the results with experiment shews that in respect of the polarising effect of the obstruction there is complete agreement, but that theory and experiment give widely different results in respect of the variation of the intensity of the waves.

The principal difference between the results obtained on the assumption that the semi-infinite plane is perfectly absorbing and on the assumption that it is perfectly conducting, is that in the first case the change of phase is a quarter of a wave length and in the second case one-eighth of a wave length. The effect of this will be that the positions of the diffraction bands will be different in the two cases, and, if their positions could be observed accurately, the nature of the condition to be satisfied at the surface of the plane, assuming that the effect of the plane can be represented by a linear relation to be satisfied at its surface, could be deduced. It is, however, more probable that the effect of a metallic screen cannot be so represented, but that there is a thin layer at its surface throughout which the equation to be satisfied gradually changes.

The above results can be applied to shew how an approximation to the effect of an open end in a condenser formed by parallel plane conducting plates can be obtained. Let the plane $y = 0$ be that which is midway between the two plates, the planes of the plates being $y = h$, $y = -h$, their edges lying in the plane $x = 0$, and the axis of z being parallel to their edges; then, considering the waves in the space between the plates, they give rise to waves propagated from the open end, the directions of the incident waves coming along the condenser being those which lie between $-\pi/2$ and $\pi/2$, the angle being

* Poincaré, *l. c.* p. 187.

measured from the axis of x. It therefore follows that at some
distance from the edges of the plates the plane $x = 0$ forms the
boundary of the shadow, and hence it may be assumed that the
effect of the open end to the left of the plane $x = 0$ outside the
condenser is negligible. If c denotes the magnetic induction,
which is everywhere parallel to the axis of z, then c satisfies
the same conditions as the velocity potential of the wave
motion of air in the space would satisfy; hence the problem to
be solved is analytically identical with the acoustical one, and
the method of approximation used by Lord Rayleigh[*] can be
adopted. The magnetic induction to the right of the plane
$x = 0$ is given by

$$c = -\frac{1}{2\pi} \int_{-h}^{h} K_0 (\iota \kappa R) \, dy',$$

where

$$R^2 = x^2 + (y - y')^2,$$

and $2\pi/\kappa$ is the wave length, it being assumed that $\dfrac{\partial c}{\partial x} = 1$
over the open end.

In order to calculate the value of

$$\int_{-h}^{h} c \, dy,$$

it will be further assumed that the distance apart of the plates
of the condenser is small compared with the wave length;
hence in calculating the value of c to be used in the above
expression it can be assumed that κR is small, and therefore

$$c = -\frac{1}{\pi} \int_{-h}^{h} \left[\Pi' (0) - \frac{\pi \iota}{2} - \log \frac{\kappa R}{2} \right] dy',$$

that is,

$$c = \frac{2h}{\pi} \left\{ \log \frac{\kappa}{2} - \Pi' (0) \right\} + h\iota + \frac{1}{\pi} \int_{-h}^{h} \log R \, dy',$$

whence over the open end

$$c = \frac{2h}{\pi} \left\{ \log \frac{\kappa}{2} - \Pi' (0) - 1 \right\} + h\iota + \frac{1}{\pi} (h + y) \log (h + y)$$
$$+ \frac{1}{\pi} (h - y) \log (h - y).$$

[*] *Theory of Sound,* vol. II. §§ 307, 312.

Substituting this value of c and performing the integrations it follows that

$$\int_{-h}^{h} c\,dy = 2h \left[\frac{2h}{\pi} \left\{ \log \kappa h - \Pi'(0) - \frac{3}{2} \right\} + h\iota \right].$$

To the left of the plane $x = 0$, that is, between the plates of the condenser the value of c may be written

$$c = [A \cos \kappa x + B \sin \kappa x]\, e^{\iota n t},$$

where the relation between the constants is given by

$$\frac{A}{\kappa B} = \frac{2h}{\pi} \left\{ \log \kappa h - \Pi'(0) - \frac{3}{2} \right\} + h\iota ;$$

whence, putting $B = 1$,

$$c = \left[\left\{ \frac{2\kappa h}{\pi} \left(\log \kappa h - \Pi'(0) - \frac{3}{2} \right) + h\iota \right\} \cos \kappa x + \sin \kappa x \right] e^{\iota n t},$$

which, taking the real part, gives

$$c = \sin \kappa\,(x - \alpha) \cos nt - \kappa h \cos \kappa x \sin nt,$$

where α, which is small, is the distance of the nearest loop from the open end and is given by

$$\alpha = \frac{2h}{\pi} \left[\log \frac{\lambda}{2\pi h} + \Pi'(0) + \frac{3}{2} \right],$$

λ denoting the wave length. This is the result stated in § 37.

INDEX.

The numbers refer to pages.

Printed in the United States
By Bookmasters